T0302292

MICROSCOPY HANDBOOKS 48

# Electron Energy Loss Spectroscopy

# Royal Microscopical Society MICROSCOPY HANDBOOKS

## Series Advisors

Angela Kohler (Life Sciences), *Alfred Wegener Institut, Notke-Strasse 85, 22607 Hamburg, Germany*

Mark Rainforth (Materials Sciences), *Department of Engineering Materials, University of Sheffield, Sheffield S1 3JD, UK*

# Electron Energy Loss Spectroscopy

**Rik Brydson**

LEMAS Centre, Department of Materials, School of Process, Environmental and Materials Engineering, University of Leeds, Leeds, UK

Taylor & Francis
Taylor & Francis Group

LONDON AND NEW YORK

© Taylor & Francis, 2001

First published 2001

All rights reserved. No part of this book may be reproduced or transmitted, in any form or by any means, without permission.

A CIP catalogue record for this book is available from the British Library.

ISBN 1 85996 134 7

**Published by Taylor & Francis**
**2 Park Square, Milton Park, Abingdon, Oxon, OX14 4RN**
**270 Madison Ave, New York NY 10016**

Transferred to Digital Printing 2006

Production Editor: Paul Barlass.
Typeset by Marksbury Multimedia Ltd, Midsomer Norton, Bath, UK.

**Front cover:** RGB processed image obtained from three EFTEM elemental maps of the interface between a boride-coated fibre and a titanium alloy matrix in a fibre reinforced metal matrix composite. Red, titanium; green, boron; blue, gadolinium. Image reproduced with permission from Brydson R, Hofer F, *et al.*, *Micron* 1996; **27:** 107–120. Image recorded at FELMI-TUGRAZ by F. Hofer.

**Publisher's Note**
The publisher has gone to great lengths to ensure the quality of this reprint but points out that some imperfections in the original may be apparent

# Contents

# Abbreviations

| | |
|---|---|
| ADC | analogue to digital converter |
| ADF | annular dark field |
| AES | Auger electron spectroscopy |
| ALCHEMI | atom location by channelling-enhanced microanalysis |
| AO | atomic orbital |
| BF | bright field |
| BIS | Bremstrahlung isochromat spectroscopy |
| CCD | charge coupled device |
| CL | cathodoluminescence |
| CTEM | conventional TEM |
| DF | dark field |
| DFT | density functional theory |
| DOS | density of states |
| DQE | detector quantum efficiency |
| ECOSS | electron Compton scattering from solids |
| EDX | energy dispersive X-ray |
| EELS | electron energy loss spectroscopy |
| EFTEM | energy-filtered transmission electron microscopy |
| ELNES | electron energy loss near-edge structure |
| EPMA | electron microprobe analyser |
| ESI | electron spectroscopic imaging |
| EXAFS | extended X-ray absorption fine structure |
| EXELFS | extended electron energy loss fine structure |
| FIB | focused ion beam |
| FIM | field ion microscopy |
| FWHM | full width half maximum |
| GIF | Gatan imaging filter |
| GOS | generalized oscillator strength |
| HAADF | high-angle annular dark field |
| HREELS | high resolution electron energy loss spectroscopy |
| IPS | inverse photoemission spectroscopy |
| IR | infrared |
| JDOS | joint density of states |
| LAMMA | laser microprobe mass analysis |
| LDA | local density approximation |
| MASNMR | magic angle spinning nuclear magnetic resonance |
| MO | molecular orbital |

| | |
|---|---|
| MS | multiple scattering |
| MSA | multivariate statistical analysis |
| MSR | multiple scattering resonance |
| PDA | photodiode array |
| PEELS | parallel electron energy loss spectroscopy |
| PIXE | proton induced X-ray emission |
| POSAP | position sensitive atom probe |
| PSF | point spread function |
| RBS | Rutherford backscattering spectrometry |
| RDF | radial distribution function |
| RGB | red/green/blue |
| SAED | selected area electron diffraction |
| SBR | signal to background ratio |
| SCF | self consistent field |
| SEA | spectrometer entrance aperture |
| SEELS | surface electron energy loss spectroscopy |
| SEM | scanning electron microscopy |
| SIMS | secondary ion mass spectrometry |
| SNMS | secondary neutral mass spectrometry |
| SNR | signal-to-noise ratio |
| SSD | single scattering distribution |
| STEM | scanning transmission electron microscopy |
| TEM | transmission electron microscopy |
| UHV | ultra high vacuum |
| UPS | ultraviolet photoelectron spectroscopy |
| UV | ultraviolet |
| WDX | wavelength dispersive X-ray |
| XANES | X-ray absorption near-edge structure |
| XAS | X-ray absorption spectroscopy |
| XPS | X-ray photoelectron spectroscopy |
| ZL(P) | zero loss (peak) |

# Preface

Imaging in the transmission electron microscope (TEM) or scanning TEM (STEM) relies on the transmission of high-energy electrons through extremely thin regions of bulk materials. However, imaging solely intensity or phase variations of the electron beam after transmission through a sample can only provide a certain degree of information. Potentially, the transmitted beam contains a wealth of additional spectroscopic information arising from inelastic scattering, and hence energy loss processes, of the incident beam via interaction with atoms within the sample. Parallel electron energy loss spectroscopy (PEELS) and energy-filtered transmission electron microscopic (EFTEM) imaging in the environment of the TEM relies on the interrogation of these transmitted electron energies and provides an increasingly important analytical tool for the detection and location of elements at high spatial resolution. A further benefit of electron spectroscopy in the TEM is the ability to determine local electronic structures for the extraction of the local chemical bonding from a spatially distinct region within a microstructure such as a second phase particle, an interface or a defect.

EELS in the TEM has rapidly become an established method for the characterization of both organic and inorganic materials at the microscopic level. With the increasing interest in performing science and technology at the dimensions of the nanoscale, EELS looks set to command an important position in the battery of techniques available to the modern-day analytical scientist. It is the purpose of this brief volume to highlight the main experimental and theoretical aspects of EELS, both qualitatively and quantitatively, and to assess the relationship of the technique to other complementary forms of analysis available in the research laboratory.

Rik Brydson

# Acknowledgements

I would like to acknowledge numerous friends and colleagues for their help and encouragement which has led to this volume, including Sam, John, Brenda, staff at BIOS, staff and students at Leeds (particularly Tony, John, Chris, David, Andy, Andrew, Howard and Clair), many long-standing collaborators throughout the UK, Europe (especially at MPI Berlin, TU Graz, MPI Stuttgart, CNRS Toulouse and CNRS Paris) and the USA and, finally, Old Holborn and Timothy Taylor for their continual support.

# Safety

Attention to safety aspects is an integral part of all laboratory procedures and both the Health and Safety at Work Act and the COSHH regulations impose legal requirements on those persons planning or carrying out such procedures.

In this and other Handbooks every effort has been made to ensure that the recipes, formulae and practical procedures are accurate and safe. However, it remains the responsibility of the reader to ensure that the procedures which are followed are carried out in a safe manner and that all necessary COSHH requirements have been looked-up and implemented. Any specific safety instructions relating to items of laboratory equipment must also be followed.

# 1 Introduction

## 1.1 What is EELS?

Considering the large range of possible physical analysis techniques available, very few microscopically analytical techniques are currently in widespread use. There are some 10 or more possible primary probes (e.g. electrons, X-rays, ions, atoms, light, neutrons, sound etc.) which can be used to excite up to 10 secondary effects from the region of interest (e.g. electrons, X-rays, ions, light, neutrons, sound, heat etc.); the chosen secondary effects may then be monitored as a function of one or more of seven or eight variables (e.g. energy, temperature, mass, intensity, time, angle, phase etc.) as well as a function of position in the sample. This gives the possibility of around 700 single signal techniques as well as the technically more difficult multi-signal techniques. At present the number of techniques which have been tried is around 100. In this book we will mainly be concerned with those analytical techniques which are based in the electron microscope and hence we only consider those in which the primary probe is an electron beam.

Electron energy loss spectroscopy (EELS) is the general title and acronym for techniques whereby a beam of electrons is allowed to interact with matter (in principle, a solid, liquid or gaseous specimen) and the scattered beam of electrons is spectroscopically analysed to give the energy spectrum of electrons after the interaction. The EELS spectrum so formed can also be filtered and the filtered electrons recombined to form an image of the specimen – a technique known as electron spectroscopic imaging (ESI). In general terms this set of techniques may be classed as an absorption spectroscopy since energy and intensity from the incident electron beam is absorbed by the matter under study. As we shall see, the nature of this energy absorption will depend on the precise composition and electronic structure of the matter under investigation.

The subject of this book is the technique of EELS conducted in the environment of a transmission electron microscope (TEM), whereby a focused electron beam is allowed to traverse a thin solid sample (although gases are also sometimes studied) and the transmitted electron beam is spectroscopically analysed. Hence this may be termed *transmission EELS* and is known by the acronyms EELS or parallel EELS (PEELS) – the associated spectroscopic imaging mode being known as energy filtered TEM (EFTEM) or ESI. An alternative technique also known as EELS, or high-resolution EELS (HREELS), or even surface EELS (SEELS) is concerned solely with the interaction of an electron beam with a surface of a material with the backscattered electron beam being analysed. These surface specific techniques allow the study of surface adsorbates and surface layers; however, they will not concern us here and for further information the reader is referred to the book by Ibach and Mills (1982).

## 1.2 Interaction of electrons with matter

The use of electrons as a probe of the atomic and electronic structure of solids has many advantages. Firstly, electrons are relatively easy to produce via simple heating of say a tungsten filament. Secondly, there is a large interaction between the incident electron beam and the electrons in the solid, which may be used for the purpose of analysis. Finally, since electrons are charged particles they may be accelerated to high kinetic energies via the application of a high voltage and subsequently focused onto small areas using electromagnetic lenses so forming images or diffraction patterns at atomic-scale resolution.

### 1.2.1 Electronic structure of atoms and solids

A simplified picture of the structure of an isolated atom is shown in *Figure 1.1*. The dense positively charged nucleus is exactly neutralized by the surrounding negative electrons, which describe orbits around this central nucleus; these orbits are shown in *Figure 1.1* as simple spherical Bohr shells. The set of electronic wavefunctions and associated electronic energy levels for the simple case of an isolated hydrogen atom, which consists of one proton and one electron, may be obtained by solving the Schrödinger equation for a single electron moving in the potential of the Hydrogen nucleus (Levine, 1991). The wavefunctions, $\Psi$, may be expressed as a *radial part*, governing the spatial extent of the wavefunction, multiplied by a *spherical harmonic*, which determines the exact shape. These localized atomic states are known as Rydberg states and may be described in terms of either simple Bohr shells or as combinations of the three quantum numbers $n$, $l$ and $m$, known as electron orbitals. The Bohr shells (designated $K$, $L$, $M$ etc.) correspond to the principal quantum numbers ($n$) equal to 1, 2, 3 *et cetera*. Within each of these shells, the electrons may exist in $s$, $p$, $d$, or $f$ subshells, for which the angular-momentum quantum number ($l$) equals 0, 1, 2, 3, respectively. It

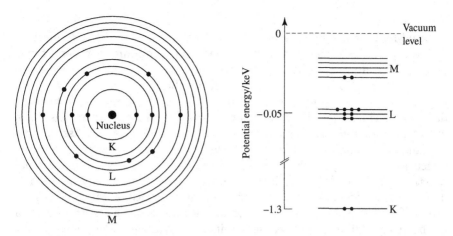

**Figure 1.1.** The Bohr shell electronic structure and associated energy level diagram of an isolated magnesium atom.

is usual to define the zero-of-energy scale (the vacuum level) as the potential energy of a free electron far from the atom. The energies of the localized electrons are then negative (i.e. they are bound to the atom) as shown in *Figure 1.1*. *Table 1.1* gives the '*KLM*' and '*spdf*' descriptions of the 16 lowest electron energy states in the atom, together with the number of electrons which each state can hold. The occupation of the states depends on the total number of electrons in the atom. In the hydrogen atom, which contains only one electron, the set of Rydberg states is almost entirely empty except for the $1s$ level which is half full. The energy spacing between these states becomes smaller and smaller and eventually converges to a value known as the vacuum level ($n=\infty$) which corresponds to the ionization of the inner-shell electron. Above this energy the electron is free of the atom and this is represented by a

**Table 1.1.** Electron states in the atom including the standard nomenclature for the EELS ionization edge involving transitions from these initial states

| Bohr shell | *KLM* subshell description | *spdf (j)* description | Occupancy | Final state symmetry after dipole transition |
|---|---|---|---|---|
| *K* | *K* | $1s$ ($1s_{1/2}$) | 2 | *p* |
| *L* | $L_1$ | $2s$ ($2s_{1/2}$) | 2 | *p* |
| | $L_2$ | $2p$ ($2p_{1/2}$) | 2 | *s* or *d* |
| | $L_3$ | $2p$ ($2p_{3/2}$) | 4 | *s* or *d* |
| *M* | $M_1$ | $3s$ ($3s_{1/2}$) | 2 | *p* |
| | $M_2$ | $3p$ ($3p_{1/2}$) | 2 | *s* or *d* |
| | $M_3$ | $3p$ ($3p_{3/2}$) | 4 | *s* or *d* |
| | $M_4$ | $3d$ ($3d_{3/2}$) | 4 | *p* or *f* |
| | $M_5$ | $3d$ ($3d_{5/2}$) | 6 | *p* or *f* |
| *N* | $N_1$ | $4s$ ($4s_{1/2}$) | 2 | *p* |
| | $N_2$ | $4p$ ($4p_{1/2}$) | 2 | *s* or *d* |
| | $N_3$ | $4p$ ($4p_{3/2}$) | 4 | *s* or *d* |
| | $N_4$ | $4d$ ($4d_{3/2}$) | 4 | *p* or *f* |
| | $N_5$ | $4d$ ($4d_{5/2}$) | 6 | *p* or *f* |
| | $N_6$ | $4f$ ($4f_{5/2}$) | 6 | *d* or *g* |
| | $N_7$ | $4f$ ($4f_{7/2}$) | 8 | *d* or *g* |

continuum of empty states. In fact the critical energy to ionize a single isolated hydrogen atom is equal to 13.61 eV and is termed a Rydberg.

Solving the Schrödinger equation for higher atomic number atoms, with more than one electron, is difficult due to the problem of electron–electron interactions in the expression for the potential. However, one method is to assume that the wavefunctions have a similar form to those derived for the simple H atom (known as hydrogenic-like solutions). The wavefunctions and energies are then derived self-consistently. Assuming a starting potential, initial solutions of the Schrödinger equation are derived which are, in turn, fed back to construct a new and improved potential function. The calculation is iterated until a minimum-energy solution is derived. This approach is known as the variation principle. The final results are self-consistent field (SCF) atomic wavefunctions and they have a qualitatively similar form to those obtained for hydrogen, which leads to the basis for the periodic classification of the elements in the Periodic Table.

When atoms come into close proximity with other atoms in a solid, most of the electrons remain localized and may be considered to remain associated with a particular atom, however, some outer electrons will become involved in bonding with neighbouring atoms. Upon bonding the atomic energy level diagram in *Figure 1.1* becomes modified (Cox, 1991). Briefly, the well-defined outer electron states of the atom overlap with those on neighbouring atoms and become broadened into energy bands. One convenient way of picturing this is to envisage the solid as large molecule. *Figure 1.2* shows the effect of increasing the number of atoms on the electronic energy levels of a one-dimensional solid (i.e. a linear chain of atoms). For a simple diatomic molecule, the two outermost atomic orbitals (AOs) overlap to produce two molecular orbitals (MOs) which can be viewed as a linear combination of the two atomic orbitals.

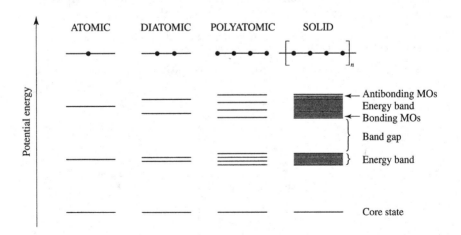

**Figure 1.2.** Schematic representation of the electron energy level diagram on going from an isolated atom to a gradually larger linear chain of *n* atoms. As *n* increases the number of molecular orbitals correspondingly increases and the outer energy levels progressively become more closely spaced until they form an overlapping energy band.

One MO, formed from the in-phase overlap of the AOs, is lower in energy than the corresponding AOs and is termed a bonding MO, whilst the other MO formed from the out of phase overlap is higher in energy than the corresponding AOs and is termed an antibonding MO. In principle, this overlap can be either parallel or perpendicular to the internuclear axis – known as σ and π bonding, respectively. Gradually increasing the length of the chain increases the total number of MOs and these gradually overlap to form bands of allowed energy levels, which are separated by forbidden energy regions arising from the original energy gaps between the atomic orbitals of the isolated atoms. The degree to which the orbitals are concentrated at a particular energy is reflected in the *density of states*, $N(E)$, where $N(E)\, dE$ is the number of allowed energy levels per unit volume of the solid in the energy range $E$ to $E+dE$.

An alternative picture for the origin of the energy band structure of solids arises from the simple Drude–Lorentz model for metallic solids, which pictures a metal as an array of positive ions in a sea or gas of free valence electrons (Kittel, 1996). Here the wave-like nature of the electron is considered which leads to only certain electron wavelengths, λ (or alternatively wavevectors $k=2\,\pi/\lambda$), and hence energies, being allowed to exist within the physical dimensions of the metal volume. In fact, for bulk materials, these allowed energy levels are extremely close together forming a continuous band of allowed energies. However, in crystalline solids, diffraction of the electron waves occurs at certain critical energies resulting in standing waves and associated forbidden energy regions, which disrupt this continuous band structure. The total number of valence electrons in the solid fill the allowed energy bands, each individual electronic level accommodating two electrons of opposite spin. The highest occupied level is known as the Fermi level, $E_F$.

The outermost bands in a solid are known as the valence and conduction bands. In a metal the conduction band is unfilled ($E_F$ lies within a band) and contains the 'free' electrons, which are responsible for the conduction and much of the bonding. In an insulator or semiconductor the valence band is full, while the conduction band, separated from the valence band by a forbidden energy gap (the band gap), remains empty; $E_F$ lies within the band gap.

As we shall see in Chapters 2, 4, 5 and 6, EELS involves the excitation of electrons in both the localized atomic-like levels and the occupied energy bands to higher lying unoccupied energy levels. EELS in transmission mode, therefore allows us to probe the electronic structure of bulk solids and, if performed in an electron microscope, this can be achieved at extremely high spatial resolution, at best approaching that of atomic dimensions.

## 1.2.2 *The electron as an analytical probe*

Outer electrons are easily removed from an atom or a solid since only a small amount of energy need be supplied either in the form of heat or under the application of a strong electric field. These two techniques for the production electrons for the purposes of analysis are known as thermionic and field emission respectively.

Considering an electron as a particle, it carries a single negative charge, $e$, of $1.6 \times 10^{-19}$ C and has a rest mass, $m_e$, of roughly $0.9 \times 10^{-30}$ kg. If a single electron is accelerated by a large voltage, $V$, then its velocity, $v$, can approach the velocity of light, $c$, and relativistic effects need to be considered. One such effect is that the mass will increase

$$m = m_e/(1-(v/c)^2)^{1/2} = \gamma m_e \qquad (1.1)$$

where $\gamma$ is known as the relativistic factor and is equal to $(1-(v/c)^2)^{-1/2}$. Considering the electron as a wave, the wavelength, $\lambda$, is given by the de Broglie relationship as $\lambda = h/(mv)$, where $h$ is Planck's constant. The energy given to the electron during acceleration, $eV$, is related to the energy represented by the relativistic change in mass:

$$eV = (m - m_e)c^2 \qquad (1.2)$$

Thus the wavelength of the electron may be related to the potential difference or accelerating voltage, $V_0$:

$$\lambda \text{ (in nm)} = h^2/(2eV_0 m_e + e^2 V_0^2/c^2) = [1.5/(V_0 + 10^{-6} V_0^2)]^{1/2} \qquad (1.3)$$

If relativistic effects are ignored, Equation (1.3) reduces to $\lambda = (1.5/V_0)^{1/2}$ nm. At the accelerating voltages which are most useful for electron microscopy (20 kV and above) the electrons are accelerated to a velocity which is a significant fraction of the velocity of light and relativistic effects are then important. *Table 1.2* demonstrates the effect of relativity for some common accelerating voltages employed in the TEM.

### 1.2.3 *The scattering of electrons by atoms*

In the majority of electron microscopes, high-energy primary electrons are incident on the specimen and, either the same electrons, or electrons produced via a variety of secondary effects exit the specimen to form an image or a measured signal. Consequently it is vitally important to

**Table 1.2.** Table of electron wavelengths for common accelerating voltages in the TEM

| keV | $\lambda$ (non-relativistic)/nm | $\lambda$ (relativistic)/nm | Comments |
|---|---|---|---|
| 97 | 0.00393 | 0.00375 | 100 keV GIF |
| 100 | 0.00387 | 0.00370 | 100 keV PEELS |
| 117 | 0.00358 | 0.00339 | 120 keV GIF |
| 120 | 0.00354 | 0.00335 | 120 keV PEELS |
| 197 | 0.00276 | 0.00252 | 200 keV GIF |
| 200 | 0.00274 | 0.00251 | 200 keV PEELS |
| 297 | 0.00225 | 0.00197 | 300 keV GIF |
| 300 | 0.00224 | 0.00197 | 300 keV PEELS |
| 397 | 0.00194 | 0.00164 | 400 keV GIF |
| 400 | 0.00194 | 0.00164 | 400 keV PEELS |
| 1000 | 0.00122 | 0.00087 | |

understand the possible interactions between high-energy electrons and the atoms of the specimen. Once this understanding has been achieved it is then possible to interpret each type of analytical signal including the EELS spectrum.

There is a set of common terminology for electron scattering. The probability that an electron will be scattered by a particular electron–specimen interaction is usually described in terms of either a cross-section, $\sigma$, or a mean free path, $\Lambda$. For a given scattering process, the cross-section represents the area which the scattering particle appears to present to the electron. If there are $N$ particles per unit volume of specimen and the cross-section for a particular scattering event is $\sigma$, then the probability of a single electron being scattered in this way during its passage through a specimen thickness $dx$ is $N\sigma dx$. An alternative way of expressing the same idea is to define the mean free path for the scattering process, $\Lambda$, as

$$\Lambda = 1/(N\sigma) \tag{1.4}$$

$\Lambda$ has the dimension of length and represents the average distance which an electron will travel before being scattered by a particular interaction in the specimen.

For the majority of scattering processes, mean free paths are similar to the thickness of a thin TEM specimen, which means that during transmission electrons are scattered either once or not at all. If an electron is incident on a thick scanning electron microscopy (SEM) specimen, it will be scattered many times before it is stopped and loses its kinetic energy. The terms single scattering, plural scattering and multiple scattering are used to describe situations in which electrons are scattered once, several times and many times, respectively.

For a particular scattering process, the probability of an electron suffering $n$ scattering events, $p(n)$, whilst travelling a distance $x$ through a material, is given by Poisson statistics:

$$p(n)=(1/n!)\,(x/\Lambda)^n\,\exp(-x/\Lambda) \tag{1.5}$$

When many scattering events, with possibly different mechanisms, occur, averaging approaches such as computer-based Monte-Carlo methods are more useful for analysing electron scattering in materials (Materials Science on CD-ROM, 1998).

During scattering both the amplitude and phase of the incident electron wave may alter. There are two main types of scattering processes known as *elastic* and *inelastic* scattering.

***Elastic scattering.*** Elastic scattering is generally coherent (i.e. the phase relationship between scattered waves from different atoms is preserved) and involves no change in the energy of the primary electron, although there may be large changes in direction. In electron microscopy, elastic scattering is the major mechanism by which electrons are deflected and is the main contribution to diffraction patterns and images.

In the elastic scattering of electrons by atoms there are two basic types of event (Hammond, 1997).

(i)  Large-angle elastic scattering through angles greater than 5°, which is also known as Rutherford scattering. This results from the direct Coulombic interaction between the primary electron and the nucleus of the atom. This process arises when the incident electron travels very close to the nucleus and therefore the cross-section depends strongly on atomic number (effectively the size of the nucleus) and varies as $Z^2$.

(ii) Small-angle elastic scattering (typically a few degrees or 10–100 milliradians) occurs when the electron travels much farther from the nucleus and arises from the scattering of the incident electron by the screened nuclear field, the screening arising from the electron cloud of the atom.

The probability of elastic scattering is given by the square of the atomic scattering amplitude for electrons, $f$, which is a function of the scattering angle, $\theta$,

$$f(\theta) = me^2/2h^2.(\lambda/\sin\ \theta)^2.(Z-f_x) \tag{1.6}$$

The term involving $Z$ represents the Rutherford scattering from the nucleus, this term being modified by $f_x$ (the atomic scattering factor for X-rays) which then represents scattering from the screened nuclear field. The amplitude falls off rapidly at large angles and, for a given $\theta$, decreases as $\lambda$ decreases (i.e. the incident energy increases).

In crystalline solids, the low-angle scattering from the electron cloud is influenced dramatically by the periodic nature of the atomic arrangment and hence the electron distribution. The smooth $f(\theta)$ function no longer applies since we have interference effects between the coherent waves scattered from regularly spaced atomic planes and strong diffracted waves are generated in specific directions given by Bragg's Law. High-angle scattering is unaffected and depends only on the nature of the atoms and not their relative positions and follows the $f(\theta)$ function (Equation 1.6) at high angles.

The mean free path for elastic scattering depends strongly on atomic number (Egerton, 1996) and is given by

$$\sigma = 1/\Lambda = [1-Z/(596(v/c))].(1.5 \times 10^{-24}\ \mathrm{m}^2)\ Z^{3/2}/(v/c)^2 \tag{1.7}$$

For 100-keV electrons, $\Lambda$ is roughly 5 nm for gold ($Z = 79$) but increases to about 150 nm for carbon ($Z = 6$).

Although elastic scattering strictly refers to scattering involving no exchange of energy, this is really only true for small scattering angles. For a scattering angle of 180°, a head-on collision, the energy transfer may exceed 1 eV for a 100-keV incident electron and may be sufficient to displace the atom from its position in a crystalline lattice resulting in knock-on or displacement damage. Furthermore, as the scattering angle increases, the elastic scattering component becomes increasingly more incoherent in nature.

***Inelastic scattering.*** Inelastic scattering is incoherent (i.e. any phase relationships between scattered waves are lost) and involves a loss in the energy of the incident electrons. It is the energy analysis of the inelastically scattered electron beam that forms the basis for EELS. The inelastically scattered beam is generally concentrated about the incident beam direction compared to the more widely spread (although still forward-peaked) elastic scattering. The total cross-section for inelastic scattering varies approximately linearly with $Z$.

There are many interaction processes that can cause energy to be lost by the primary electron and transferred to the electrons or atoms of the specimen. Here we consider four of the most probable types of scattering event. Inelastic scattering processes are eventually responsible for the stopping of an electron by a bulk solid with almost all of the kinetic energy carried by the primary electron ending up as heat (phonons) in the specimen. A small proportion of the energy may escape as either X-rays or secondary electrons, resulting from the secondary process of de-excitation of the atoms involved in the scattering events (see section on relaxation of excited atoms, below). As well as primary effects to the incident electron beam, such as energy loss, which forms the basis for EELS, these associated secondary effects may also be employed for the purposes of analysis or imaging and are discussed in the Section 1.2.5.

The main types of inelastic scattering mechanisms are:

*(a) Phonon excitation.* The primary electron excites phonons, which are atomic vibrations in the solid, and in doing so effectively raises the local temperature. The amount of energy loss is small, generally much less than 1 eV and is generally not resolved in EELS measurements in the TEM, although phonon losses are employed in surface HREELS spectroscopy. The scattering angle is large (*ca.* 10°) and the mean free path is typically 1 µm. All electrons which remain in the solid are likely to excite phonons eventually, perhaps after they have lost larger amounts of energy (see below).

*(b) Plasmon excitation.* The primary electron excites collective, 'resonant' oscillations (plasmons) of the valence electrons in a solid, which in the simple Drude–Lorentz model can be pictured as oscillations of the electron gas. In exciting a plasmon, the primary electron typically loses between 5 and 30 eV and the mean free path is usually about 100 nm. Due to this short mean free path, plasmon scattering is the most frequent scattering process in electron–solid interactions and therefore dominates the electron energy loss (EEL) spectrum in the low energy loss region, as we shall see in Chapters 2 and 4. An additional aspect of this high probability is that plasmon scattering may complicate other inelastic signals and may need to be removed for the purposes of analysis – see Chapter 5.

*(c) Single electron excitation.* The primary electron transfers some of its energy to a single electron in the material resulting in ionization. Lightly bound valence electrons may be ejected from atoms and if they escape

from the specimen they may be used for secondary electron imaging. Energy losses for such excitations range up to typically 50 eV. If deeply bound inner shell electrons are removed the energy loss is much greater, that is, it takes 284 eV to ionize a carbon $K$ (1$s$) electron and over 2200 eV to ionize a zirconium $L$. The mean free path for this type of scattering is quite large, of the order of micrometres, so the process occurs much less frequently than plasmon scattering. However, despite this drawback these energy loss processes of the incident beam form the basis for high-energy loss EELS and the secondary effects produced when the excited or ionized atom relaxes (e.g. X-ray or Auger electron production) are also ideal for elemental analysis.

In common with most scattering processes, the cross-section for single electron excitation decreases as the primary electron energy, $E_0$, increases. The scattering cross-section also decreases as the energy of excitation (the energy loss suffered by the incident electron), $E$, increases – for a given incident energy, the cross-section varies roughly as

$$\sigma \propto N/E \tag{1.8}$$

where $N$ is the number of electrons in the particular inner shell undergoing ionization.

*(d) Direct radiation losses.* Fast incident electrons can also lose energy in solids due to deceleration processes in which energy is emitted directly in the form of photons. In this case the energy loss can be anything from zero up to the primary electron energy. This radiation, known as *bremsstrahlung* or braking radiation, forms the background radiation in the X-ray emission spectrum upon which the characteristic emitted X-rays, arising from ionization of single inner shell electrons and subsequent de-excitation, are superimposed (Hren *et al.*, 1979).

***Inelastic scattering and absorption.*** If a high-energy incident electron enters a bulk, solid specimen then all the various types of inelastic scattering processes can take place. These scattering events will occur until either the electron is stopped or it leaves from the surface by which it entered. The trajectories of the electrons may be simulated on a computer using interactive probablity (Monte-Carlo) calculations. In these calculations the electron is allowed to enter a solid specimen with a known incident energy (Materials Science on CD-ROM, 1998). Depending on the various probabilities, different types of scattering events will occur through various angles. A typical set of electron trajectories is shown in *Figure 1.3*. It may be seen that the majority of electrons are brought to rest within the solid but a few are backscattered and leave the specimen. The volume within which typically 95% of the primary electrons are stopped or absorbed is referred to as the interaction volume. The size of the interaction volume is dependent on the atomic number of the specimen and the incident beam energy.

For the case of a thin TEM sample, many of the primary electrons will be transmitted through the specimen. The electron detection system,

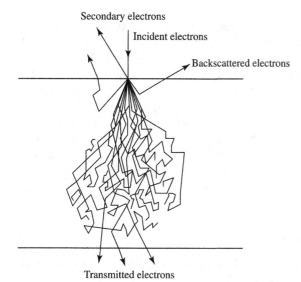

Secondary electrons

Incident electrons

Backscattered electrons

Transmitted electrons

**Figure 1.3.** Typical set of electron trajectories in a solid as produced by a Monte-Carlo calculation.

located below the specimen, will be of a finite size and will subtend a certain angle, $\theta$, at the specimen (in reality a two-dimensional detector will subtend a solid angle, $\Omega$). For transmitted electrons, the term 'absorbed' essentially means 'not detected', which in this case corresponds to 'scattered through an angle greater than $\theta$'. If the atomic cross-section for scattering through an angle $> \theta$ is $\sigma_a$, then we can describe absorption in a thin specimen using the conventional Beer–Lambert law. The fractional intensity, $I/I_O$, which remains after absorption in a thickness $x$ is given by

$$I/I_O = \exp\left(-N\sigma_a x\right) \tag{1.9}$$

where $N = N_A \rho / A$ is the number of scattering atoms per unit volume, $N_A$ is Avogadro's number, $\rho$ the density and $A$ is the atomic mass. As most scattering mechanisms, both elastic and inelastic, are strongly forward-peaked the value of the cross-section, $\sigma_a$, will therefore be a function of the angle $\theta$ subtended by the detector. We return to this point in Sections 2.4 and 5.3.

### 1.2.4 Primary and secondary signals arising from the scattering processes

When a high-energy electron beam is incident on a solid, a variety of processes occur, which have already been discussed in Section 1.2.3. In a thin specimen employed in the TEM, we are concerned with the situation where the electron beam traverses the specimen; however, essentially the same is true for reflection from a surface when the angle of incidence between the electron beam and the surface is small (glancing angle). The signals created by the electron–solid interaction are shown schematically in *Figure 1.4* and may be conveniently divided into two groups:

(i) Primary or direct processes such as the elastic and inelastic scattering of the incident electron beam and the excitation of electrons in the solid. In a crystalline solid the elastically scattered electrons are contained in the Bragg diffracted beams and may be used to form diffraction patterns as well as focused to form images of the structure. The inelastically scattered component of the scattered electron beam has undergone the processes described in the section on inelastic scattering and, if this component is dispersed according to the precise energy, it may be used to form an EEL spectrum.

(ii) Secondary processes which occur as a result of electron–electron scattering and the subsequent de-excitation of atoms in the solid (e.g. production of X-rays, Auger electrons, photons etc.).

### *1.2.5 Secondary effects*

Loosely, a secondary effect is defined as an effect occurring as a result of scattering of the primary beam, which can be detected outside the specimen. These effects are mainly the production of electrons or electromagnetic radiation shown in *Figure 1.4*. For simplicity we may categorize them into five types which are related to the way each effect is exploited in analysis.

***Secondary electrons.*** This is a general term for electrons that escape from the specimen with energies below about 50 eV. They could conceivably be primary electrons, which, at the very end of their trajectory, reach the surface with a small proportion of their energy remaining. However, they are more likely to be electrons originally bound to atoms in the sample to which a small amount of energy has been

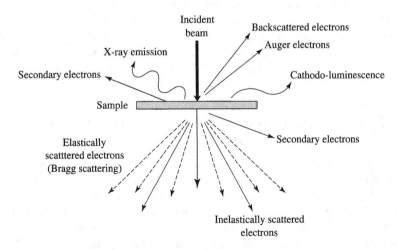

**Figure 1.4.** Schematic diagram of the signals created during transmission of a high-energy electron beam through a thin solid sample.

transferred within a short distance of the surface. Secondary electrons are very abundant, their yield, $\partial$ (the number emitted per primary electron) can reach as high as 1. As a consequence they are the most commonly employed imaging signal in SEM and provide topographic contrast from the specimen surface.

***Backscattered electrons.*** Some primary electrons undergo large deflections (Rutherford backscattering) and leave the specimen surface with the remainder of their energy intact – these are termed back-scattered electrons and have correspondingly high energies. Strictly backscattering is a primary effect and the backscattered electron yield, $\eta$, is small compared with $\partial$, however, backscattered electrons are used for imaging and chemical phase identification in the SEM since their yield is very sensitive to the mean atomic number of the specimen at the incident electron probe position.

***Relaxation of excited atoms.*** If an atomic electron has been ionized or excited to an empty higher energy level, the atom is in an excited, high-energy state. Subsequently, the empty electron state or hole will be filled by an electron dropping down from a higher occupied energy level and the atom will relax. The excess energy will be released via a secondary effect usually involving the emission of another particle or a photon of radiation. There are three main ways in which this relaxation can happen:

*(a) Cathodoluminescence.* If the vacant electron state is an outer state, possibly in the valence band of the solid, then the energy given off will be small and is commonly emitted in the form of a photon in the ultraviolet, visible or infrared part of the electromagnetic spectrum. This electron-induced luminescence is known as cathodoluminescence (CL). In an insulator or semiconductor, the CL signal will provide a local measure of the band gap energy, which may be affected by the presence of dopant atoms or structural changes.

Alternatively, if the vacant state is an inner state, the amount of energy to be released is larger and there are two main possibilities: the emission of an X-ray photon or an Auger electron, shown in *Figures 1.5a* and *1.5b*.
*(b) X-ray emission.* The energy of the X-ray photon emitted when a single outer electron drops into the inner shell hole is given by the difference between the energies of the two excited states involved. For example, if a *K*-shell electron is ionized from a molybdenum atom and is replaced by an *L* electron falling into the vacant state the energy difference, $\Delta E$, is 17 400 eV which is emitted as the $K\alpha$ X-ray of Mo; the X-ray wavelength, given by $\lambda = hc/\Delta E$, is 0.071 nm. Alternatively, the inner shell hole could be filled by an electron dropping from the higher energy Mo *M* shell which would result in the emission of an Mo $K\beta$ X-ray of energy 19 600 eV and wavelength 0.063 nm. Due to the well-defined nature of the various atomic energy levels (Section 1.2.1), it is clear that the energies and

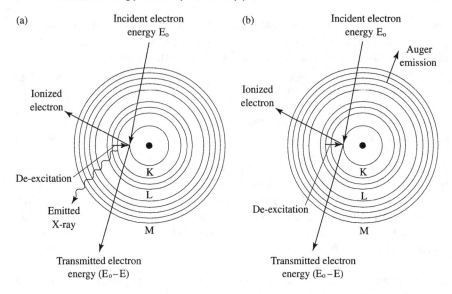

**Figure 1.5.** De-excitation mechanisms for an atom that has undergone *K*-shell ionization: (a) emission of a characteristic *K*α X-ray and (b) emission of a *KLM* Auger electron.

wavelengths of the set of emitted X-rays will have characteristic values for each of the atomic species present in the specimen. By measuring either the energies or wavelengths it is possible to determine which elements are present at the particular position of the electron probe; this is the basis for energy dispersive and wavelength dispersive X-ray analysis (EDX and WDX, respectively; Hren *et al.*, 1979; Jones, 1992) in the electron microscope.

It would appear that there would be a large set of characteristic X-rays for each atom, since transitions between all possible states listed in *Table 1.1* would appear to be allowed. However, a set of dipole selection rules prohibit many of the expected transitions such as a transition between a 2s state and a 1s state. Numerous general electron microscopy texts given in the reference section (Goodhew *et al.*, 2000; Hren *et al.*, 1979; Jones, 1992; Loretto, 1994; Williams and Carter, 1997) list the most useful X-rays for microanalysis in the TEM.

*Figure 1.6* shows a typical electron-generated X-ray emission spectrum from molybdenum oxide. The Mo *K*α, *K*β and *L* X-ray lines as well as the O *K*α line are superimposed upon a background arising from *bremstrahlung* (braking radiation). The latter X-rays are not characteristic of any particular atom but depend principally on specimen thickness.

*(c) Auger emission.* Besides X-ray emission, an alternative relaxation mechanism for excited atoms is the ejection of an outer electron, which carries off any excess energy as kinetic energy. This process is known as Auger emission and is shown in *Figure 1.5b*. Here three electrons are now involved: the original vacancy, the outer electron that fills it, and another outer electron, which is ejected with any surplus energy and possibly

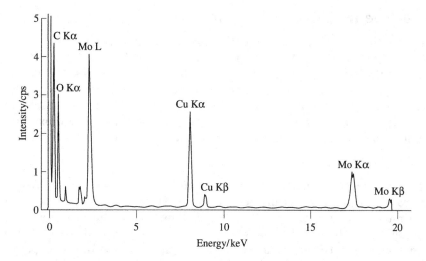

**Figure 1.6.** Schematic energy dispersive X-ray (EDX) emission spectrum from a thin specimen of molybdenum oxide on a carbon support film. The copper $K\alpha$ and $K\beta$ X-ray peaks are due to the specimen holder.

detected. The notation for Auger transitions reflects the involvement of these three energy levels in the form of a subscript, for example the oxygen Auger transition arising from $K$-shell ionization is denoted $O_{KLL}$. Measurement of the energy of characteristic Auger electrons (usually in the range 100–1000 eV) is the basis for Auger electron spectroscopy (AES) which is an important analytical tool and is extremely surface specific owing to the small escape depth of the low-energy emitted Auger electrons (Briggs and Seah, 1990a). Besides providing quantitative surface composition, in many cases, AES can also provide some chemical specificity from small shifts in the energies of the Auger peaks (Watts, 1990).

Auger electron production and X-ray emission are alternative processes for the relaxation of excited atoms. The two processes, however, do not occur with equal probability and the fraction of atoms which emit an X-ray, known as the fluorescence yield, varies strongly with atomic number, $Z$. The fluorescence yield, $\omega$, is given by

$$\omega = Z^4/(Z^4 + \text{constant}) \qquad (1.10)$$

The constant has a value of about $10^6$ for $K$-shell vacancies and is larger for $L$- and $M$-shell vacancies. For low $Z$ elements, Auger electrons will be emitted in far larger numbers than X-rays, while the reverse is true for heavy elements.

The relative abundances of the three types of electron emission (secondary, backscattered and Auger) are very different, and for bulk specimens there are large numbers of secondary and backscattered electrons (with $\partial > \eta$), but relatively few of the analytically most useful Auger electrons.

## *1.2.6 The interaction volume*

In order to assess the size and location of the area being analysed by electron–matter interactions, we need to know how far into the specimen the electron beam can penetrate and also from which regions the various signals can escape. *Figure 1.7* shows schematically the shape of the interaction volume in both solid and thin specimens. In a solid specimen of density $\rho$, the electron penetration depth, $R$, and thus the approximate diameter of the interaction volume, is found to vary as

$$R \propto E_0^q / \rho \qquad (1.11)$$

where $q$ generally lies between 1.5 and 2. As stated above, Auger electrons and more generally secondary electrons can only escape from the first few nanometres of the surface. This means that these signals originate from the top of the interaction volume, the lateral size of which is determined by the electron probe size which in turn defines the spatial resolution in AES and SEM. Backscattered electrons are higher in energy and can escape from deeper below the surface where the interaction volume is wider laterally, this results in a degraded image resolution for backscattered electron imaging in the SEM (relative to that of secondary electron imaging). X-rays of energy $E_c$ will not be excited by electrons whose energy has fallen below $E_c$ so the expression for the depth from which X-rays are excited is of the form

$$R \propto (E_0^q - E_c^q)/\rho \qquad (1.12)$$

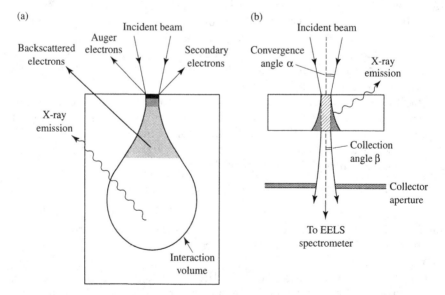

**Figure 1.7.** Schematic diagram of the shape of the interaction volume in (a) a bulk and (b) a thin specimen also showing the regions from which the different analytical signals are produced. For a thin specimen, the use of an angle-limiting EELS collector aperture significantly improves the spatial resolution over that obtained with EDX.

A commonly used set of parameters is $q=1.68$ and the constant of proportionality is around $10^{-2}$, which gives $R$ in micrometres if $E_0$ is in keV and $\rho$ in g cm$^{-3}$. Except for CL photons, soft X-rays from light elements and for some quantitative work, it is usually assumed that most X-rays generated within the interaction volume are able to escape from the specimen.

In a thin TEM specimen, penetration is of little concern; however, there may be some broadening of the beam as it passes through the specimen. For a foil of thickness $t$ (nm), density $\rho$ (g cm$^{-3}$) and atomic weight $A$, the beam broadening (in nanometres) can be approximated by

$$b = 0.198\ (\rho/A)^{1/2}.(Z/E_0).t^{3/2} \tag{1.13}$$

where $E_0$ is the incident beam energy in keV and $Z$ is the atomic number (Jones, 1992). At 100 keV a 100-nm thick sample gives a typical beam broadening of the order of 10 nm. This beam broadening will limit the ultimate (lateral) spatial resolution for microanalysis using small probes, particularly for EDX analysis where we collect all X-rays produced isotropically within the beam broadened volume. As is shown in *Figure 1.7*, in the case of EELS in transmission, where the signal is highly forward-peaked, we tend to define the signal collection angle ($\beta$) using a well-defined collector aperture which can significantly improve the ultimate spatial resolution compared to that obtained with EDX.

# 1.3 Basics of the TEM

The general subject of transmission electron microscopy is vast and has been the subject of numerous texts since the first basic TEM was designed and built by Ruska and Knoll in Berlin in the early 1930s. A comprehensive and relatively recent text is that due to Williams and Carter (1997). Here we simply summarize the basic experimental design, which has relevance to the experimental use of transmission EELS described in later chapters.

A conventional TEM (CTEM) consists of a number of electron-optical components shown schematically in *Figure 1.8*. Firstly, the electron source or gun consists of a filament, which is typically tungsten or a ceramic material with a low work function such as lanthanum hexaboride (LaB$_6$), and a focusing electrode (Wehnelt). Emission of electrons is by either thermionic emission (W or LaB$_6$) or by field emission (single crystal W) or, in recent machines, a combination of the two (a Schottky emitter). The type of electron gun determines the total emission current from the gun, the source size (these two are related by the quantity of brightness: the current per unit area) and the spread of electron energies contained in the electron beam.

The electron beam is accelerated towards the anode, typically the acceleration is in a number of stages and the final beam energy can

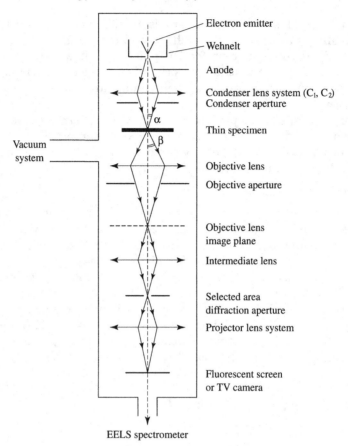

**Figure 1.8.** Schematic diagram of the basic components of a TEM.

typically be between 100 and 400 keV in most standard commercial instruments.

Two or more electromagnetic condenser lenses demagnify the probe to a size typically between a few microns and few nanometres; the excitation of these lenses controls both the beam diameter and the beam divergence/convergence. Generally the first condenser (C1 or *spot size*) controls the demagnification of the source, while the second (C2 or *intensity*) controls the size of the spot at the specimen and hence the beam divergence/convergence.

The specimen is in the form of a thin (< 100 nm) 3-mm disc of either the material itself or the material supported on an electron transparent grid. Specimen preparation techniques are discussed in Goodhew (1985). The specimen is usually inserted into the vacuum of the TEM via an airlock and fixed into a side-entry specimen rod, which can be translated or tilted (about one or two axes).

The main electromagnetic objective lens forms the first intermediate, real-space projection image of the illuminated specimen area (in the image plane of the lens) as well as the corresponding reciprocal space diffraction pattern (in the back focal plane of the lens). Here the image magnification relative to the specimen is typically 50–100 ×. The objective aperture can be inserted in the back focal plane of the objective lens to limit beam divergence in reciprocal space of the transmitted electrons contributing to the magnified image. Typically there are a number of circular objective apertures ranging from 10 to 100 microns in diameter.

The projector lens system consists of a first projector or intermediate lens which focuses on either the objective lens image plane (microscope operating in imaging mode) or the back focal plane (microscope in diffraction mode). The first projector lens is followed by a series of three or four further projector lenses – each of which magnify the image or diffraction pattern by typically up to 20 ×. The selected area electron diffraction (SAED) aperture usually lies in the image plane of one of the projector lenses (due to space considerations) and, if projected back to the first intermediate image, and hence the specimen, effectively allows the selection of a much smaller area on the specimen for the purposes of forming a diffraction pattern.

The overall microscope system gives a total magnification of up to 1 million times on the electron fluorescent microscope viewing screen, camera or (below this) the self-contained EELS spectrometer and detection system which almost always possesses a variable entrance aperture itself. A schematic of this experimental arrangement is shown in *Figure 1.9*, and is discussed in Chapter 3; here $\alpha$ and $\beta$ are known as the convergence and collection semi-angles respectively. Note that certain commercial EELS systems (particularly those initially developed for the purposes of energy filtered imaging – see Chapter 7) employ an in-column design with the EELS spectrometer placed between the objective and projector lenses.

A CTEM image is recorded in parallel and consists of a combination of mass-thickness contrast, diffraction contrast and at higher magnifications phase contrast. TEM image formation and simulation is a subject in its own right (Buseck *et al.*, 1988). Image resolution is primarily determined by imperfections or aberrations in the objective lens; two common aberrations are known as *spherical aberration* (whereby electrons from one object point, travelling through the lens at differing distances from the optic axis, are focused to different points in the image) and *chromatic aberration* (whereby electrons from one object point, which have differing energies, are focused to different points in the image).

A variant on this transmission imaging system arose after the development of the SEM; developed in 1968, the scanning TEM (STEM) serially scans a small (typically nanometre) probe, produced by an electron gun/lens system, in a two-dimensional raster across the specimen (Keyse *et al.*, 1998). At each point the transmitted beam intensity is measured and contributes to an intensity on a cathode ray tube or digitally recorded – so building up a serial image of the specimen, the resolution of which is determined primarily by the probe size on the specimen. Two intensities are usually recorded, that falling on both an on-axis bright field (BF) detector which collects electrons which have

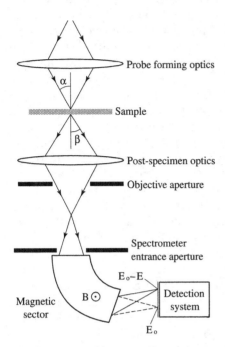

**Figure 1.9.** Schematic diagram of an EELS spectrometer attached to a CTEM; $\alpha$ and $\beta$ are known as the convergence and collection angles respectively.

undergone relatively small angles of scattering (principally undiffracted and inelastically scattered electrons) as well as an annular dark field (ADF) detector which collects higher angle Bragg diffracted electrons. Retraction of the BF detector allows electrons to enter an EELS spectrometer. A schematic diagram of a STEM is shown in *Figure 1.10*.

Dedicated STEMs employ extremely small probe sizes produced by cold field emission electron sources and can provide extremely high-spatial and high-energy resolution EELS measurements, while there are a number of hybrid TEM/STEM instruments which can operate in both modes.

Finally, *Table 1.3* presents a summary of experimental TEM-based techniques; these include what may be termed both imaging (microstructural) and analysis (compositional) techniques. The table also attempts to highlight the important additional and complementary information, which the technique of transmission EELS indirectly provides. These techniques are summarized in Chapter 2 and discussed in detail in Chapters 4–8.

## 1.4 Comparison of EELS in TEM with other spectroscopies

The relationship of EELS with other TEM-based spectroscopic techniques is outlined in *Table 1.3*. In the electron microscope, the most

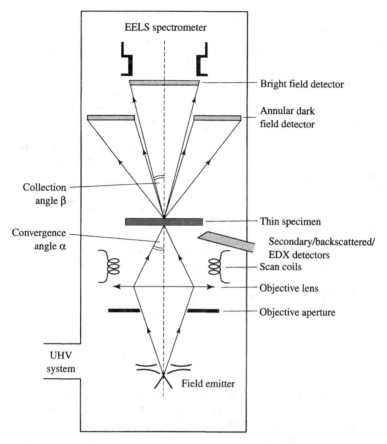

EELS spectrometer

Bright field detector

Annular dark
field detector

Collection
angle β

Convergence
angle α

Thin specimen

Secondary/backscattered/
EDX detectors

Scan coils

Objective lens

Objective aperture

UHV
system

Field emitter

**Figure 1.10**. Schematic diagram of the basic components of a dedicated STEM. Note the field emission source is usually at the bottom of the microscope column for reasons of stability.

complementary spectroscopic technique is usually X-ray emission measured by either EDX or WDX analysis. WDX is most commonly used for analysis of bulk specimens in an electron microprobe analyser (EPMA). High sensitivity (0.01 atom%) for most elements and good accuracy ($\pm$ 1 atom%) can be achieved with a spatial resolution of a few cubic microns; however, the serial nature of the WDX collection technique results in long recording times. The parallel collection (and hence more rapid), yet less sensitive lower spectral resolution technique of EDX is found in both scanning and TEM and, in the TEM, gives higher spatial resolution for analysis than in the SEM due to the reduced beam broadening in a thin specimen (Section 1.2.6).

## 1.4.1 EDX in the TEM

A specimen volume irradiated with a focused electron probe in the TEM will emit characteristic X-rays isotropically of which typical EDX detectors collect roughly 1%; this may be compared with the collection

**Table 1.3**. Summary of CTEM/STEM techniques highlighting the additional benefits of combining these with EELS

| Technique | Information provided | Additional benefits combined with EELS |
|---|---|---|
| **Imaging techniques** | | |
| CTEM, diffraction contrast | Microstructure | EFTEM<br>Phase mapping,<br>Concentration mapping |
| CTEM, phase contrast | Atomic structure | EFTEM<br>Removal of inelastic background<br>Chemically specific information to<br>inform atomic structure |
| Selected area electron diffraction | Crystal structure | Orientation-dependent EELS<br>EFTEM<br>Removal of inelastic background<br>contribution |
| Convergent beam electron diffraction | Crystal structure,<br>crystallography and<br>space group symmetry | EFTEM<br>Removal of inelastic background<br>contribution |
| STEM bright field | Microstructure/atomic<br>structure | Energy filtered BF STEM |
| STEM, high-angle annular dark field | Microstructure/atomic<br>structure/atomic number | EELS Point analysis<br>Linescans<br>Spectrum imaging |
| **Analysis** | | |
| Energy dispersive X-ray analysis | Medium + heavy element<br>bulk composition | Full quantitative microanalysis<br>across Periodic Table<br>Higher spatial resolution analysis<br>Bonding information |
| Cathodo-luminescence | Band gaps, defects | Bonding information |
| Auger electron spectroscopy | Surface composition<br>plus some chemistry | Bulk composition and bonding |

efficiency of an EELS system which may approach 50% owing to the strongly forward-peaked signal. Even in thin specimens, significant beam-broadening effects occur for nanometre electron probes, which is the limiting factor for the spatial resolution of X-ray emission analysis in the TEM. The current generation of EDX detectors employs ultra-thin windows which permit detection of elements as light as boron, although extensive absorption of low energy X-rays in the specimen may limit accurate quantification of first row elements in the Periodic Table. The energy resolution of current EDX detectors is typically 100–150 eV, which may give rise to peak overlap problems at low X-ray energies and effectively precludes any chemical state information being extracted from any changes in X-ray binding energies. However, the recent advent of high-energy resolution detectors based on microcalorimetry may alter this position in the future.

Besides the relative ease of acquisition of EDX spectra and subsequent analysis, an important advantage of EDX over EELS lies in the low background signal in the EDX spectrum which, as discussed in the section on inelastic scattering, arises from *bremsstrahlung* radiation. This results in a much higher signal-to-background ratio for EDX compared to EELS and, even though both EELS and EDX background signals increase with increasing sample thickness, EDX can generally tolerate thicker sample areas although quantification of lighter elements X-rays can then become more difficult. Accurate quantification of EDX spectra is generally based on sensitivity or $k$-factors, which need to be experimentally determined for the particular detector system, microscope conditions and specimen-detector geometry; this means EDX, unlike EELS, is not an absolute, standardless method of quantification.

A more detailed comparison between EELS and EDX quantification in an analytical TEM or STEM is given in Section 5.7. However, EELS has interesting analogues in other spectroscopic techniques based on the use of both incident photons and ions. Many of these techniques provide complementary information to EELS at different levels of both lateral and depth resolution and can often be used together to provide a full microstructural description of a specimen.

## 1.4.2 Photon-based techniques

In terms of the energy range accessible to experimental measurement, EELS overlaps the soft X-ray, ultraviolet (UV) and to some extent optical energy regions of the electromagnetic spectrum. Considering the close relationship between X-ray diffraction (XRD) and electron diffraction for the determination of structural information, the most directly relevant technique to transmission EELS is X-ray absorption spectroscopy (XAS), usually measured in transmission in an X-ray synchrotron beamline; in particular the specific techniques X-ray absorption near-edge structure (XANES) and extended X-ray absorption fine structure (EXAFS) which probe the absorbed X-ray intensity due to inner shell ionization events (Fuggle and Inglesfield, 1992; Stohr, 1992). The use of incident photons currently limits the spatial resolution of the spectroscopic data due to the problems in focusing the incident beam to a size comparable with electron probes in the TEM. However, at the time of writing, the use of zone plates does allow the focusing of soft X-rays to probe sizes of the order of 50 nm, resulting in X-ray microscopy based on XAS that has some advantages in terms of reduced specimen radiation damage, the ability to conduct *in situ* experiments and the possibility of extracting three-dimensional tomographic information on elemental and bonding distributions within a sample.

Another photon beam-based technique is the surface specific X-ray or UV photoelectron spectroscopy (XPS/UPS) which uses incident X-rays or UV radiation to excite surface photoelectrons (Briggs and Seah, 1990a). In a similar fashion to XPS, EELS can provide elemental quantification as well as bonding information; however, this is from the bulk not the surface. While XPS records simple inner shell ionization events (i.e. electrons excited to states free of the atom and the solid), EELS includes energy transitions to bound electron states within the solid (as in XAS), which complicates the spectral interpretation. Low-energy-loss EELS data is analogous to UV photoelectron spectroscopy (UPS) and optical absorption or reflectivity data. Photoelectron microscopy (PEEM) is also possible by immersing the sample in a strong magnetic field and imaging photolectrons of specific energies from a particular surface region typically from between 10 and 0.1 microns in dimension.

Inverse photoemission spectroscopic (IPS) techniques, including *Bremsstrahlung* Isochromat Spectroscopy (BIS), use a near-monochromatic beam of electrons which relax into unoccupied electronic states of the solid with the simultaneous emission of X-rays or UV photons, dependent on the electron energy. In BIS, the electron energy is varied and the emission of photons at one particular wavelength is monitored; alternatively, in IPS, the electron energy is generally held constant and the spectrum of emitted photons is recorded. Both techniques provide a means of probing directly the unoccupied electron states of a solid in a similar fashion to EELS and in particular electron energy loss near-edge structure (ELNES) techniques. However, unlike EELS, the spectroscopic mechanism in IPS is not element specific; it is generally surface sensitive and not a transmission-based technique and, finally, it does not involve the creation of a core hole during the excitation process, which can provide a useful comparison to the results of ELNES studies (Chapter 6).

Other regions of the electromagnetic spectrum also provide extremely useful data for characterizing bonding in a sample, particularly infra-red (IR) and Raman techniques which rely on the excitation of phonons within the sample; again spatial resolution is generally much poorer than electron-based techniques.

Magic-angle spinning nuclear magnetic resonance (MASNMR) relies on the absorption of radio frequency radiation to excite nuclei between magnetic spin states; this is a powerful tool for determining the local atomic environments in a sample can complement EELS data in structural characterization of materials (Richardson *et al.*, 1993).

### *1.4.3 Ion-based techniques*

Techniques such as secondary ion mass spectrometry (SIMS) and secondary neutral mass spectrometry (SNMS) rely on the sputtering of either charged or neutral atomic or molecular surface fragments by an incident ion beam and their analysis by a mass spectrometer (Briggs and Seah, 1990b). They are essentially destructive analytical techniques used to determine composition, a related photon-based technique being laser microprobe mass analysis (LAMMA). Although these techniques can detect virtually all elements in the periodic table with a high sensitivity (typically ppm), spectra are difficult to quantify accurately (with perhaps the exception of SNMS) due to the complicated nature of the sputtering and ionization process and the interaction with the surrounding matrix material. Typical lateral spatial resolutions obtainable with focused ion or laser probes are between 0.1 and 1 micron and although, especially in the case of SIMS and SNMS, the analysis is extremely surface sensitive, continuous acquisition of spectra during analysis allows a depth profile of elemental distributions to be determined.

The use of extremely high-energy ions of low mass (e.g. $H^+$ or $He^{2+}$) produced in an accelerator results in considerable elastic backscattering of the incident ion flux. Analysis of the energy loss spectrum of the backscattered ions is the basis for Rutherford backscattering spectroscopy (RBS), the spectrum exhibiting edges at energies which are characteristic of each backscattering element present and whose widths are dependent on the elemental depth distribution. Although the lateral spatial resolution of the technique is poor, RBS does provide a non-destructive means of three-dimensional elemental analysis. A related technique is proton-induced X-ray emission (PIXE), which allows the determination of low elemental concentrations (Chu *et al.*, 1978).

Atom probe field ion microscopy (FIM) involves the removal of atoms from a sharply pointed, conducting specimen tip, typically a few nanometres in diameter, by a strong electric field. The desorbed atoms or ions can be chemically identified using a time of flight mass spectrometer giving extremely high spatial resolution maps of elemental distributions. A recent varient, the position sensitive atom probe (POSAP), allows a three-dimensional reconstruction of the atomic structure in the tip (Miller, 2000).

## 1.5 Conclusions

In this first chapter we have briefly covered some of the background material assumed in the later chapters. Firstly, we have summarized the electronic structure of the atom and indicated how this is influenced by the presence of neighbouring atoms in a solid. We have discussed how transmission EELS arises from the scattering of high-energy electrons by atoms in a solid; usually these electrons are produced in a TEM and we have highlighted the various operating modes of these instruments and summarized the benefits and additional information which EELS provides. Finally we have related this information to that provided by other commonly used analytical techniques.

## References

**Briggs, D. and Seah, M.P.** (1990a) *Practical Surface Analysis, Volume 1: Auger and X-ray Photoelectron Spectroscopy.* Wiley, Chichester.

**Briggs, D. and Seah, M.P.** (1990b) *Practical Surface Analysis, Volume 2: Ion and Neutral Mass Spectrometry.* Wiley, Chichester.

**Buseck, P., Cowley, J. and Eyring, L. (eds)** (1988) *High Resolution Transmission Electron Microscopy and Associated Techniques.* Oxford University Press, Oxford.

**Chu, W.K., Mayer, J.W. and Nicolet, M.A.** (1978) *Backscattering Spectrometry.* Academic Press, New York.

**Cox, P.A.** (1991) *The Electronic Structure and Chemistry of Solids.* Oxford Science, Oxford.

**Egerton, R.F.** (1996) *Electron Energy Loss Spectroscopy in the Electron Microscope.* Plenum Press, New York.

**Fuggle, J.C. and Inglesfield, J.E. (eds)** (1992) *Unoccupied Electronic states. Fundamentals for XANES, EELS, IPS and BIS.* Springer Verlag, Berlin.

**Goodhew, P.J.** (1985) Thin foil preparation for electron microscopy. In: A.M. Glauert (ed). *Practical Methods in Electron Microscopy*, vol. 11. Elsevier, Amsterdam.

**Goodhew, P.J., Humphreys, F.J. and Beanland, R.** (2000) *Electron Microscopy and Analysis*, 2nd Edn. Taylor and Francis, London (2000).

**Hammond, C.** (1997) *The Basics of Crystallography and Diffraction.* Oxford Science, Oxford.

**Hren, J.J., Goldstein, J.I. and Joy, D.C.** (1979) *Introduction to Analytical Electron Microscopy.* Plenum Press, New York.

**Ibach, H. and Mills, D.L.** (1982) *Electron Energy Loss Spectroscopy and Surface Vibrations.* Academic Press, New York.

**Jones, I.P.** (1992) *Chemical Microanalysis using Electron Beams.* Institute of Materials, London.

**Keyse, R.J., Garrett-Reed, A.J., Goodhew, P.J. and Lorimer, G.W.** (1998) *Introduction to Scanning Transmission Electron Microscopy.* BIOS: Oxford.

**Kittel, C.** (1996) *Introduction to Solid State Physics.* John Wiley, New York.

**Levine, I.N.** (1991) *Quantum Chemistry.* Prentice Hall, New Jersey.

**Loretto, M.H.** (1994) *Electron Beam Analysis of Materials.* Chapman and Hall, London.

**Materials Science on CD-ROM** (1998) Chapman and Hall, London. http://www.matter.org.uk

**Miller, M.K.** (2000) *Atom Probe Tomography.* Kluwer, Dordrecht (2000).

**Richardson, I.G., Brough, A., Brydson, R., Groves, G.W. and Dobson, C.M.** (1993) The Location of aluminium in substituted calcium silicate hydrate gels as determined by $^{29}$Si and $^{27}$Al NMR and EELS. *J. Am. Ceram. Soc.* **76:** 2285–2288.

**Stohr, J.** (1992) *NEXAFS Spectroscopy.* Springer Verlag, Berlin.

**Watts, J.F.** (1990) *Introduction to Surface Analysis by Electron Spectroscopy.* Oxford Science, Oxford.

**Williams, D.B. and Carter, C.B.** (1997) *Transmission Electron Microscopy.* Plenum Press, New York.

# 2 The EEL spectrum

## 2.1 The primary transmitted electron signal

Concentrating on the primary signals created by the interaction between electrons and matter discussed in Section 1.2.4, we may divide the scattered incident electrons into those which have undergone elastic collisions and lost no energy, and those which have undergone inelastic collisions with electrons in the solid and lost a small proportion of their energy. In a crystalline solid, the elastically scattered electrons are contained in the Bragg diffracted beams and may be used to form diffraction patterns as well as focused to form images of the structure. Meanwhile, the inelastically scattered component contains all the information on the electron distributions in the solid and, if the scattered electrons are dispersed according to their precise energy, they may be used to form an EEL spectrum.

## 2.2 Historical development

Historically, since the discovery of the electron by J.J. Thompson in 1897, the study of both the energy and angular distribution of electrons resulting from their scattering by matter has been extensive. The subject has been an integral part of the continued development of both scattering physics and quantum mechanics throughout the last century. Most experiments concentrated on using a near monoenergetic beam of electrons impinging on either a thin foil (transmission mode) or a thicker specimen (reflection mode) and involved measurement of either the distribution in scattering angle and/or the distribution in energy of the transmitted or reflected beam.

Early reflection-mode EELS experiments were reported by Rudberg in 1930, whilst the first measurement of the energy spectrum of transmitted electrons was performed by Ruthemann in 1941. Hillier and Baker (1944) were the first to perform transmission EELS in the environment of a specially designed TEM in 1944 and to observe extensively inner shell losses rather than low energy losses.

In 1949 the development of an electrostatic energy analyser by Möllenstadt, which could be fitted as an add-on to a normal TEM, permitted the high-resolution recording of EEL spectra. Later developments included the use a Wien filter involving a combination of electric and magnetic fields to disperse the transmitted electrons. From the 1970s onwards, EELS finally began to realise its potential as a means for the analysis of light elements at high spatial resolution via use of a simple single-prism magnetic spectrometer, which deflected the electrons through 90° and so dispersed them as a function of energy. This spectrometer was mounted beneath a conventional TEM (Marton, 1946) with the spectrometer object plane being the TEM projector lens crossover (Wittry, 1969), as shown in *Figure 1.9*. Krivanek of the Gatan company successfully designed, built and marketed a serial recording EELS system (model 607) based on a small magnetic prism spectrometer, energy selecting slit, deflection system and single-channel electron detector system (Krivanek and Swann, 1981). Alongside these developments, data processing packages designed specifically for EELS were also launched and much of the theory of EELS spectroscopy was condensed in the seminal book by Egerton (1996). In addition to the use of EELS on conventional TEM instruments, the design and marketing of the dedicated field emission STEM by Vacuum Generators in 1974 provided the vanguard for high-energy resolution and high spatial resolution EELS measurements.

The 1980s saw a large improvement in detection systems with the development of parallel position sensitive detectors based on photodiode or charge coupled diode or device (CCD) arrays. Detection methods generally relied upon projection (using quadrupole lenses) and conversion of electrons to photons on a phosphor or YAG screen with optical coupling to the diode array. Gatan launched their parallel EELS detector (model 666) in 1986 (Krivanek *et al.*, 1987), which revolutionized the technique, permitting much lower recording times (and hence lower sample drift, contamination and irradiation damage during measurement) to achieve the same statistical accuracy in inner shell EELS data as that obtained with comparable serial detection systems.

In parallel with these developments in EEL spectroscopy, the related technique of energy-filtered imaging has emerged as a powerful analytical technique in its own right allowing the removal of inelastic contrast in diffraction-based images, as well as the direct imaging of plasmon peak positions, elemental distributions and even chemical states. In 1962, Castaing and Henry (Castaing *et al.*, 1967) developed a filter based on a magnetic prism, an electrostatic mirror electrode (to keep the beam on the optical axis after energy dispersion) and an energy selecting slit. This was eventually marketed by Zeiss in an 80-kV TEM (model EM902). Subsequently a combination of a number of magnetic prisms (the omega filter; Krahl *et al.*, 1978) replaced the electrostatic mirror which had to be held at the microscope potential. This design was marketed by Zeiss (model EM912) as a 120-kV in-column energy-filtering microscope. In the early 1990s, Gatan took their magnetic prism spectrometer, corrected aberrations using quadrupoles and sextupole lenses, added an energy selecting slit and employed a two-dimensional

CCD detector array to form the Gatan imaging filter (GIF) which is essentially an add-on attachment to a conventional TEM (Krivanek *et al.*, 1995). Finally, continued developments in acquisition and data processing software (as well as data storage) have facilitated the alternative mapping procedure of spectrum imaging, where the beam is scanned and an EELS spectrum recorded at each point; post acquisition processing again permits the formation of maps of thickness, plasmon peak position or elemental concentration *et cetera* (Jeanguillaume and Colliex, 1989).

## 2.3 Basic components of an EEL spectrum

The general form of the various energy losses observed in a typical transmission EEL spectrum is shown schematically in *Figure 2.1*. This

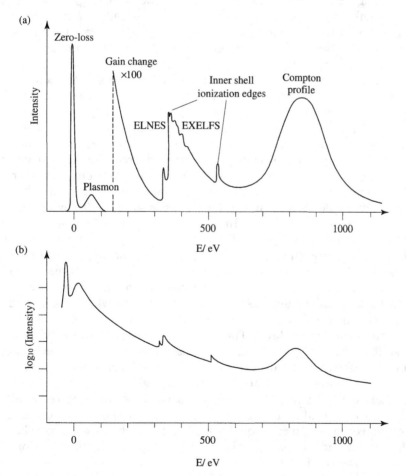

**Figure 2.1.** Schematic diagram of an EEL spectrum on both (a) a linear intensity scale and (b) a logarithmic intensity scale.

shows the scattered electron intensity as a function of the decrease in kinetic energy (energy loss, $E$) of the transmitted fast electrons and essentially represents the response of the electrons in the solid to the disturbance introduced by these incident electrons.

In a specimen of thickness less than the mean free path for inelastic scattering (roughly 100 nm at 100 keV), by far the most intense feature is the *zero loss peak* (ZLP) at 0 eV energy loss which contains all the elastically and quasi-elastically scattered electron components. Neglecting the effect of the spectrometer, the full width half-maximum (FHWM) of the ZLP is usually limited by the energy spread inherent in the electron source. In the TEM, the experimental configuration on which we will concentrate, the energy spread will generally lie between a few tenths of an eV and a few eV, depending on the type of emitter. Effectively this parameter often determines the overall resolution of the spectrum. The smallest energy losses arise from the excitation of *phonons* in the solid and these are typically in the range 10–100 meV (Chapter 1, Inelastic scattering). However, in conventional transmission EELS these are not usually resolved and are subsumed in the ZLP. However, such losses may be revealed in high-resolution reflection mode experiments where the interaction of adsorbates with surfaces may be studied – this is in essence a vibrational spectroscopic technique similar to infrared or Raman spectroscopy.

The *low loss region*, extending from 0 to about 50 eV, corresponds to the excitation of electrons in the outermost atomic orbitals, which are often delocalized (due to interatomic bonding) and extend over several atomic sites. This region therefore reflects the solid state character of the sample. The low loss region is dominated by collective, resonant oscillations of the valence electrons known as *plasmons* (Chapter 1, Inelastic scattering). The energy of the plasmon peak is governed by the density of the valence electrons in the sample and the width by the rate of decay of the resonant mode. In a thicker specimen, there would be additional peaks at multiples of the plasmon energy, corresponding to the excitation of more than one plasmon by the incident electron. The intensities of these plasmons follow the Poisson distribution discussed in Sections 1.2.3 and 4.1. A further feature in the low loss spectra of insulators are peaks known as *interband transitions*, which correspond to the excitation of valence electrons to low-energy unoccupied electronic states above the Fermi level. These single electron excitations may lead to a shift in the energy of the plasmon resonance. Generally, the low loss region is used mainly to determine the specimen thickness and to correct for the effects of multiple inelastic scattering when performing quantitative microanalysis on thicker specimens. In a more detailed analysis, the overall shape of the low loss region may be related to the dielectric response function of the material, which allows a correlation with optical measurements, including band gap determination in insulators and semiconductors.

Beyond the low loss region, the *high loss region* extends from about 40 or 50 eV to several thousand eV and corresponds to the excitation of electrons from well-localized orbitals on a single atomic site to unoccupied electron energy levels just above the Fermi level of the material. This region therefore reflects the atomic character of the specimen. As the energy loss progressively increases, this region exhibits steps or *edges* superimposed on the monotonically decreasing background electron intensity; the intensity at 2000 eV is typically eight orders of magnitude less than that at the ZLP and therefore, for clarity, in *Figure 2.1(a)* a gain change has been inserted in the linear intensity scale at 150 eV. These edges correspond to excitation of inner shell electrons (Chapter 1, Inelastic scattering) and are known as *ionization edges*. Since the energy of the edge threshold is determined by the binding energy of the particular electron subshell within an atom, which is a characteristic quantity, then the atomic type may be identified. The signal under the ionization edge extends beyond the threshold, since the amount of kinetic energy given to the excited electron is not fixed. The intensity under the edge is proportional to the number of atoms present and hence this allows the technique to be used for quantitative analysis as discussed in Chapter 5. The various edges are classified using the standard spectroscopic notation, e.g. $K$ excitation for the ionization of $1s$ electrons, $L_1$ for $2s$, $L_2$ for $2p_{1/2}$, $L_3$ for $2p_{3/2}$, $M_1$ for $3s$ *et cetera*. The subscript refers to the total angular momentum quantum number, $j$, of the electron which is equal to the orbital angular momentum, $l$, plus the spin quantum number, $s$. For a $2p$ electron, $l = 1$, and this can couple to the spin of the electron in one of two ways, that is, $j = l+s = 1+1/2 = 3/2$ $(L_3)$ or $j = 1 - 1/2 = 1/2$ $(L_2)$. *Table 1.1* in Chapter 1 summarizes the nomenclature commonly used in the classification of EELS ionization edges (Ahn and Krivanek, 1983). If the sample thickness is greater than the mean free path for inelastic scattering, then multiple inelastic scattering will occur. This will increase the plasmon intensity, leading to an increase in the background contribution making it difficult to identify the presence of edges in a spectrum, as well as transferring intensity away from the edge threshold to higher energy losses due to the increased probability of double scattering events involving a plasmon excitation followed by an ionization event or vice versa.

If electrons are scattered via inelastic collisions with $K$-shell electrons of free atoms (e.g. gases) the core loss edges are sharp, saw-tooth like steps displaying no features. Other core-loss excitations in free atoms display a variety of basic edge shapes, which are essentially determined by the degree of overlap between the initial- and final-state wavefunctions which may be determined by quantum mechanics. *Figure 6.1* in Chapter 6 summarizes these commonly observed edge shapes. In solids, however, the unoccupied electronic states near the Fermi level may be appreciably modified by chemical bonding leading to a complex density of states (DOS) and this is reflected in the ELNES which modifies the basic atomic shape within the first 30–40 eV above the edge threshold. The ELNES, discussed in Chapter 6, gives information on the local structure and bonding associated with a particular atomic species, although the exact interpretation is somewhat complex. Beyond the near-edge region, superimposed on the gradually decreasing tail of the core loss edge, one observes a region of weaker, extended oscillations known as the extended energy loss fine structure (EXELFS). The period of the oscillations may be used to determine bond distances, while the amplitude reflects the coordination number of the particular atom. In terms of scattering theory, the main distinction between the ELNES and EXELFS regions lies in the fact that the low kinetic energy of the ejected electron in the near-edge region means that it samples a greater volume (the inelastic mean free path is large), and multiple elastic scattering occurs so providing geometrical information. In the EXELFS regime, the higher kinetic energy results in predominantly single elastic scattering of the ejected electron so giving short-range information.

If the transmitted electrons are collected over small scattering angles (corresponding to small momentum transfers upon collision), then the dipole selection rules apply to the various transitions observed at the core loss edges, as is the case for X-ray emission. This limits the observed electron transitions to those in which the angular momentum quantum number, $l$, changes by $\pm 1$. This results in different edges of the same element probing different symmetries of the final state, that is, a $K$-edge will probe the unoccupied $p$-like DOS, whereas an $L_{2,3}$-edge will probe the unoccupied $s$- and $d$-like DOS. At larger scattering angles the dipole selection rules break down and other transitions are observed. At very large angles, a new regime is encountered in which the electrons in the sample may be regarded as if they were free and a hard sphere collision occurs with an associated large momentum transfer resulting in the *Compton profile*. The width of this feature represents a Doppler broadening of the energy of the scattered electrons due to the initial state momentum of the electrons in the sample and can, in principle, give bonding information as highlighted in Chapter 8.

## 2.4 Basic physics

For the various scattering processes described in Chapter 1, it is common to represent the scattering of fast incident electrons by atoms in the sample in terms of a vector diagram shown in *Figure 2.2*. Before scattering, the incident electrons will have a particular energy, $E_0=(h/2\pi)^2 k_0^2/(2m_e)$, wavelength, $\lambda_0$ (see Equation 1.3 where $V=V_0=E_0/e$) and hence wave vector $\mathbf{k}_0=2\pi/\lambda_0$; $\mathbf{k}$ is similar to frequency as it is an inverse wavelength but note this a vector quantity and has both magnitude (the length of the vector, equal to $|\mathbf{k}|$) and direction. Note the incident beam energy, $E_0 = (h/2\pi)^2 k_0^2/(2m_e)$. Experimentally the range of incident wavevectors will be defined by the convergence semi-angle, $\alpha$, in the microscope (*Figure 1.9*).

After scattering, the wavevector will have changed to $\mathbf{k}_f$ (the range of wavevectors collected will depend on the collection semi-angle, $\beta$ – see

*Figure 1.9* and Chapter 3); this represents a change in both direction and magnitude for an inelastic scattering process (*Figure 2.2a*), whereas for an elastic process, $|\mathbf{k_0}|=|\mathbf{k_f}|$ as shown in *Figure 2.2b*. The momentum transfer suffered by the incident electron is given by $(h/2\pi).\mathbf{q}$ where $\mathbf{q}=\mathbf{k_0}-\mathbf{k_f}$, whereas the momentum transfer imparted to the specimen is correspondingly $-(h/2\pi).\mathbf{q}$.

For the elastic case, conservation of momentum gives $|q|=2|\mathbf{k}|\sin(\theta/2)$ (i.e. $q$ depends solely on scattering angle). For high incident beam energies, the momentum transfer $\mathbf{q}$ is much smaller than $\mathbf{k_0}$, and in the inelastic case, conservation of momentum and energy leads to the result

$$q^2 = k_0{}^2(\theta^2+\theta_E{}^2) \tag{2.1}$$

where $\theta$ is the scattering semi-angle and $\theta_E$ is a *characteristic* angle of scattering corresponding to the mean energy loss $E_{av}$, $\theta_E = E_{av}/(\gamma m_0 v^2)$, where $\gamma$ is the relativistic factor for a particle of velocity, $v$ is given, as before, by $\gamma=(1-v^2/c^2)^{-1/2}$ (Equation 1.1) and $m_0$ is the electron rest mass. For low beam energies (*ca.* 100 keV), $\theta_E$ approximates to $\theta_E = E_{av}/2E_0$. Values of $\theta_E$ will vary from around 10 mrad for a high energy loss edge (2000 eV) at an incident beam energy of 100 keV, to around 0.3 mrad for an edge at 200 eV at an incident beam energy of 400 keV. As discussed in Chapter 3, experimentally it is important to collect the majority of this scattering if adequate signals are to be obtained.

Since $\mathbf{q}$ is much smaller than $\mathbf{k_0}$, it is possible to decompose $\mathbf{q}$ into a component $q_{||}$ (parallel to $\mathbf{k_0}$) and $q_{\perp}$ (perpendicular to $\mathbf{k_0}$) where $q^2= q_{||}{}^2+ q_{\perp}{}^2$. These are given by $q_{||}= k_0\theta_E$ and $q_{\perp}=k_0\sin\theta \approx k_0\theta$ (for small $\theta$) in accord with Equation (2.1). This has implications for orientation dependent measurements in anisotropic materials as discussed in Chapter 8.

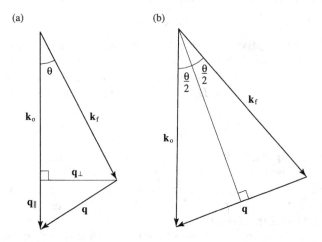

**Figure 2.2.** Wavevector diagram for (a) inelastic and (b) elastic scattering. For the inelastic case, the decomposition of the momentum transfer into components parallel and perpendicular to the incident beam direction is also shown.

What is experimentally measured in an EELS spectrum is related to the total inelastic cross-section for scattering, $\sigma$, or more exactly its dependence on energy loss, $E$, and solid scattering angle, $\Omega$, known as the double differential cross-section, $d^2\sigma/(dE\ d\Omega)$. This quantity represents the fraction of incident electrons (energy $E_0$) which are scattered into a solid angle $d\Omega$ ($=2\pi \sin\theta\ d\theta$, if we ignore any diffraction effects which may destroy the cylindrical symmetry of the experimental measurement obtained with a circular collection aperture) with an energy of between $E$ and $E+dE$.

The form of the double differential cross-section can be derived using quantum mechanics via a number of equivalent approaches. One description, known as *Fermi's Golden Rule*, relies on time-dependent perturbation theory to describe the transition rate between the initial and final states of the interacting electrons.

It is assumed that the kinetic energies of the incident electrons are much greater than the energies of the excited atomic states – this is known as the Born approximation. This allows the total wavefunction of the system to be written as a product of the atomic states and plane wave states. The plane wave states describe the fast incident electron before and after scattering and are of the general form $\exp(i\mathbf{k}.\mathbf{r}')$, where $k$ is the wavevector (either $\mathbf{k_0}$ or $\mathbf{k_f}$) and $\mathbf{r}'$ the vector coordinate describing the position of the fast electron. The fast electron interacts through the Coulomb potential (given by $V_{coul}=e^2/(4\pi\varepsilon_0\ |\mathbf{r}'-\mathbf{r}|)$) with the atomic electron which has a coordinate system $\mathbf{r}$ and causes it to make the transition from an initial atomic state wavefunction, $\Psi_i\ (\mathbf{r})$ to a final atomic state $\Psi_f\ (\mathbf{r})$. Mathematically the result is in the form of a differential cross-section with respect to solid angle given by theory due to Bethe (1964).

$$d\sigma/d\Omega = (2\pi m_e/h^2)^2(k_f/k_0)\ |\ \int\int \Psi_f{}^* \exp(i(\mathbf{k_0}-\mathbf{k_f}).\mathbf{r}')\ V_{coul}\ \Psi_i\ d^3\mathbf{r}'\ d^3\mathbf{r}|^2 \quad (2.2)$$

Integration over the coordinates of the fast electron gives

$$d\sigma/d\Omega = 4\gamma^2/(a_0{}^2\ q^4)\ (k_f/k_0)\ |\int \Psi_f{}^* \exp(i\mathbf{q}.\mathbf{r})\ \Psi_i\ d^3\mathbf{r}|^2 \quad (2.3)$$

where $\gamma$ is the relativistic correction factor (Equation 1.1) and $a_0 = \varepsilon_0 h/(e^2 m_e)$ is the Bohr radius $= 0.053$ nm – the radius of a 1s orbital in a hydrogen atom. If the final states of the atomic electron are normalized with respect to energy loss, $E$, to give $\Psi_f'\ (\mathbf{r})$, this gives us the double differential cross-section

$$d^2\sigma/(dE\ d\Omega) = 4\gamma^2/(a_0{}^2\ q^4)\ (k_f/k_0)|<f|\exp(i\mathbf{q}.\mathbf{r})|i>|^2 \quad (2.4)$$

where $<f|\ \exp(i\mathbf{q}.\mathbf{r})|i>$ denotes $\int \Psi_f'{}^* \exp(i\mathbf{q}.\mathbf{r})\ \Psi_i\ d^3\mathbf{r}$. Actually since in a solid there a number of possible (empty) band-like final states above the Fermi level, the integral term in Equation (2.4) should actually be summed over transitions to all possible final states as outlined in Equation (5.4) and in Chapter 6.

The magnitude of $\mathbf{q}=|\mathbf{q}|$ can be written in terms of the scattering angle, $\theta$, since by the cosine rule in *Figure 2.2b*, $q^2=k_0{}^2+k_f{}^2-2k_0 k_f\cos\theta$. Differentiating this with respect to $\theta$ allows us to write the solid angle in terms of $q$,

$$d\Omega = \int d\phi \; \sin\theta \; d\theta = \int d\phi \; q \; dq/k_0 k_f$$

where $\phi$ is the azimuthal angle (varying between 0 and $2\pi$ for a circular collection aperture) relative to the polar angle $\theta$. This gives

$$d^2\sigma/(dE \; dq) = 4\gamma^2 R/(Ek_0^2) \int d\phi \; (1/q) \; df/dE \qquad (2.5)$$

where $E$ is the energy loss and $R$ is the Rydberg $= (e^4 m_e/2)/(2\varepsilon_0 h)^2 = 13.61$ eV – the energy of an electron in a hydrogen $1s$ orbital. The term $df/dE$ is an atom-dependent quantity and is known as the generalized oscillator strength (GOS) per unit energy loss. This quantity depends both on energy loss, $E$, and the change in momentum, $q$,

$$df/dE = E/(Ra_0^2 q^2) \mid <f\mid \exp(i\mathbf{q.r})\mid i> \mid^2. \qquad (2.6)$$

Using Equation (2.1), we can express Equation (2.4) in terms of a double differential cross-section with respect to energy loss and scattering angle

$$d^2\sigma/(dE \; d\theta) = 4\gamma^2 \; R/(Ek_0^2) \int d\phi \; \theta/(\theta^2 + \theta_E^2) \; df/dE \qquad (2.7)$$

or alternatively in terms of solid angle as given by Equation (2.4)

$$d^2\sigma/(dE \; d\Omega) = 4\gamma^2 \; R/(Ek_0^2) \; 1/(\theta^2 + \theta_E^2) \; df/dE \qquad (2.8)$$

Roughly translated, what Equation (2.8) means is that for a particular inelastic process, such as ionization of inner-shell electrons, the angular distribution of inelastically scattered electrons will follow a Lorentzian distribution (i.e. of the form $1/(x^2 + a^2)$) of half width equal to the characteristic angle, $\theta_E$.

The theory summarized above was originally formulated by Bethe. As in X-ray scattering (Hammond, 1997), the Born approximation allows the differential cross-section to be separated into the product of two terms: one dependent on the incident electron and the other dependent only on the excited atom. These are known as an *amplitude factor* and a *dynamic structure (or inelastic form) factor* respectively and are most apparent in Equation (2.4). The amplitude factor represents the scattering of the electron by a free electron and is given by the Rutherford cross-section which is equal to $4\gamma^2/(a_0^2 \; q^4)$, while the dynamic structure factor depends on the properties of the atom and reflects the fact that the electrons are involved in bonding.

As stated, the general scattering properties of a particular atomic species are summarized in the GOS, $f$, which for a solid tends to be formulated as an energy dependent quantity, $df/dE$. A three-dimensional plot of $df/dE$ as a function of both energy and momentum supplied to the atom is known as a *Bethe surface*. The form of the GOS for the carbon $K$-shell calculated using Hydrogenic wavefunctions is shown in *Figure 2.3*. Note the presence of a maximum (Bethe ridge) at high values of momentum transfer. This represents the classical transfer of momentum to a free electron in an inelastic collision and is the Compton profile mentioned in Section 2.3 and Chapter 8.

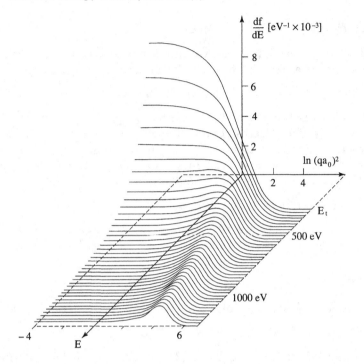

**Figure 2.3.** The Bethe surface, the GOS, $df/dE$, as a function of momentum transfer, $q$, and energy loss, $E$, for the carbon $K$-edge calculated using Hydrogenic wavefunctions. Redrawn from Egerton (1996) *Electron Energy Loss Spectroscopy in the Electron Microscope*, 2nd Edn, with permission of Plenum Publishers.

As mentioned in Section 2.3, if we use a small collection aperture and select only those scattered electrons which have undergone small momentum transfers, then the observed electronic transitions obey the dipole selection rules formulated from photon spectroscopic techniques. This is discussed in more detail in Chapter 6. However, for the present we simply note that this will affect the expansion of the exponential term in Equations (2.3) and (2.4), leading to only certain final states being selected by the excited electron. A further approximation, which is sometimes employed in the dipole regime, is that the GOS term, $df/dE$, becomes independent of $q$.

We shall return to this theoretical description in Chapters 4, 5 and 6 where we discuss aspects of quantitative spectral interpretation in EELS.

## 2.5 Summary of analytical uses

*Table 2.1*, after Disko *et al.* (1992), summarizes the analytical uses of various regions of the EEL spectrum together with the required processing steps and possible limitations. Many of these

**Table 2.1.** Summary of the analytical uses of EELS

| Property determined | Spectral region | Processing required | Uses/limitations |
| --- | --- | --- | --- |
| Sample thickness | Full spectrum or low loss region | Integration of intensities | Rapid measure of relative thickness; need Λ (inelastic) for absolute values. |
| Valence electron density | Low loss – plasmon peak position | Peak fitting if required | Phase Identification/ Effects of alloying |
| Surface/interface states | Low loss – surface plasmons | Need model of dielectric function based on geometry | Surface properties |
| Joint DOS | Low loss – energy loss function | Deconvolution/fitting of ZLP/Kramers Kronig analysis/sum rules | Band gaps/interband transitions/comparison with optical results |
| Elemental concentration/ elemental distributions (EFTEM) | High loss – ionization edge intensities | Background subtraction/ deconvolution/ integration/calculation of partial cross-sections | Good for light elements with edges in range 0–2.5 keV/ often high detection limits/limited to thin sample areas |
| Element specific local coordination and/or valency | High loss – ELNES | Deconvolution/ comparison with reference materials – fingerprinting/peak fitting | Need good energy resolution and determination of absolute energy loss |
| Element specific unoccupied DOS | High loss – ELNES | Deconvolution/modelling of electronic structure | Need good energy resolution/effect of the core hole |
| Element specific radial distribution function | High loss – EXELFS | Standard EXAFS data processing | Low SNR – requires good statistics and large edge separations |
| Anisotropic DOS | High loss – ELNES/ low loss | Deconvolution/modelling of electronic structure | Need small collection angles and well-defined specimen orientation |
| Site specific DOS | High loss – ELNES/ALCHEMI | Deconvolution/modelling of electronic structure | Need well-defined diffraction and collection conditions |
| Ground state electron momentum distribution at large momentum transfer | Compton profile | Standard Compton scattering data processing | Low SNR/diffraction conditions |

techniques are discussed in detail in subsequent chapters, however, for the present they form a useful aide-memoire for future reference.

# References

**Ahn, C.C. and Krivanek, O.L.** (1983) *EELS Atlas*. ASU Centre for Solid State Science, Tempe, Arizona and Gatan Inc., Warrendale, Pennsylvania.

**Bethe, H.A.** (1964) *Intermediate Quantum Mechanics*, 1st Edn, Chapter 15. W.A. Benjamin, New York.

**Castaing, R., Hennequin, J.F., Henry, L. and Slodzian, G.** (1967) The magnetic prism as an optical system. In: *Focussing of Charged Particles*. Academic Press, New York.

**Disko, M.M., Ahn, C.C. and Fultz, B. (eds)** (1992) *Transmission Electron Energy Loss Spectrometry in Materials Science*. TMS, Warrendale, Pennsylvania.

**Egerton, R.F.** (1996) *Electron Energy Loss Spectroscopy in the Electron Microscope*. Plenum Press, New York.

**Hammond, C.** (1997) *The Basics of Crystallography and Diffraction*. Oxford Science: Oxford.

**Hillier, J. and Baker, R.F.** (1944) Microanalysis by means of electrons. *J. Appl. Phys.* **15:** 663–675.

**Jeanguillaume, C. and Colliex, C.** (1989) Spectrum image: the next step in EELS digital acquisition and processing. *Ultramicroscopy* **28:** 252.

**Krahl, D., Herrmann, K.H. and Kunath, W.** (1978) Electron optical experiments with a magnetic imaging filter. In: *Electron Microscopy*, Vol. 1. Microscopy Society of Canada, Toronto, pp. 42–43.

**Krivanek, O.L. and Swann, P.R.** (1981) An advanced electron energy loss spectrometer. In: *Quantitative Microanalysis with High Spatial Resolution*. The Metals Society, London.

**Krivanek, O.L., Ahn, C.C. and Keeney, R.B.** (1987) Parallel detection electron spectrometer using quadrupole lenses. *Ultramicroscopy* **22:** 103–116.

**Krivanek, O.L., Friedman, S.L., Gubbens, A.J. and Krauss, B.** (1995) An imaging filter for biological applications. *Ultramicroscopy* **59:** 267–282.

**Marton, L.** (1946) Electron microscopy. *Rep. Prog. Phys.* **10:** 205–252.

**Rudberg, E.** (1930) Characteristic energy losses of electrons scattered from incandescent solids. *Proc. R. Soc. London* **A127:** 111–140.

**Ruthemann, G.** (1941) Diskrete Energieverluste schneller Elektronen in Festkorpern. *Naturwissenschaften* **29:** 648.

**Wittry, D.B.** (1969) An electron spectrometer for use with the transmission electron microscope. *Br. J. Appl. Phys. (J. Phys. D)* **2:** 1757–1766.

## *General texts on EELS* ·

**Ahn, C.C. and Krivanek, O.L.** (1983) *EELS Atlas*. ASU Centre for Solid State Science, Tempe, Arizona and Gatan Inc., Warrendale, Pennsylvania.

**Colliex, C.** (1984) Electron energy loss spectroscopy in the electron microscope. In: *Advances in Optical and Electron Microscopy 9* (eds R. Barer and V.E. Cosslett). Academic Press, London.

**Disko, M.M., Ahn, C.C. and Fultz, B. (eds)** (1992) *Transmission Electron Energy Loss Spectrometry in Materials Science*. TMS, Warrendale, Pennsylvania.

**Egerton, R.F.** (1996) *Electron Energy Loss Spectroscopy in the Electron Microscope*. Plenum Press, New York.

**Reimer, L. (ed.)** (1995) *Energy Filtering Transmission Electron Microscopy*. Springer, Heidelberg.

**Williams, B.G.** (1987) Electron energy loss spectroscopy. *Progr. Solid State Chem.* **17:** 87.

# 3 EELS instrumentation and experimental aspects

EELS originated in the environment of the TEM and this is the experimental configuration on which we will concentrate; however, various other possibilities exist, including SEELS, conducted in reflection mode. This provides the opportunity for the characterization of surface layers and complements many other surface spectroscopies. EELS conducted in transmission mode essentially probes bulk electronic and chemical properties of sample regions typically 100 nm thick. In the TEM or STEM environment it is possible to use highly focused electron probes and obtain signals at an ultimate spatial resolution of a few nanometres or less. Thus we have a very powerful technique to relate electronic and structural information with variations in morphology and composition in a microstructurally inhomogeneous specimen.

Transmission EELS involves the measurement of the distribution in energy and scattering angle of high-energy electrons (typically between 100 and 400 keV) following their passage through a thin sample. The basic experimental configuration of both TEM and STEM has been outlined in Section 1.3. Here we discuss the experimental system for measurement of EEL spectra, which is shown schematically in *Figure 1.9*.

## 3.1 The electron spectrometer

There are many types of electron spectrometer which have been developed over the years: the *sector magnet spectrometer*, the *Castaing–Henry prism*, the *Ω filter*, the *Wien filter* and the *Möllenstedt analyser* being examples of those which have been used in conjunction with TEMs, as discussed briefly in Section 2.2. The Castaing–Henry prism is commercially available but is an integral part of the microscope column and is mainly employed in dedicated energy-filtered imaging TEMs. The sector magnet spectrometer can be added to an existing TEM column and is therefore considerably more common. Hence, the following discussion is limited to sector magnets, although more detail concerning in-column filters may be found in Egerton (1996) and Reimer (1995).

A typical sector magnet consists of a homogeneous magnetic field, $B$, normal to the electron beam. This causes electrons of a given momentum, and hence kinetic energy and velocity, $v$, to follow trajectories that are essentially arcs of circles. Usually for mechanical convenience the electrons are bent through an angle of $90°$ and the magnetic force, $F=Bev$, causes electrons of different kinetic energy (and therefore energy loss) to emerge from the spectrometer spatially dispersed. In fact the spectrometer has a focussing action which results in the formation of a line spectrum (each energy loss separately focused) at the spectrometer image plane. The object plane of the spectrometer in STEM is simply the specimen or virtual image of the specimen, while in TEM it is the final projector crossover just (e.g. 40 cm) above the microscope viewing screen.

The spectrometer magnetic field is provided by two parallel pole pieces with two current carrying coils providing the excitation. The coils are symmetrically arranged about the plane midway between the pole pieces. This maintains the symmetry plane close to which the electrons travel. The return path for the magnetic flux is normally provided by using a window frame construction such that the coils are placed between the pole pieces and the gap around the edge is filled with a magnetic material. Such construction minimizes hysteresis effects and provides good shielding from external magnetic fields. In the region where the electrons enter and leave the spectrometer, the gap is left empty and the coils are displaced symmetrically from the mid-plane. In practice the pole pieces and coils are at atmospheric pressure, while an evacuated drift or flight tube (connected to the microscope vacuum system) passes between the pole pieces. An offset voltage may be applied to this drift tube which will change the kinetic energy of all the electrons and rigidly shift the whole energy loss spectrum across the detector system.

The magnetic sector spectrometer, like any other optical component, can be described by its first-order properties and its aberrations. *Figure 3.1* shows the first-order focusing properties of a sector magnet. A point source is in general brought to two line focii, one being normal to the symmetry plane and the other being in the symmetry plane. By careful choice of the angles between the electron beam and the entrance and exit faces of the magnet, these two line focii can be brought together giving a point image. Such a spectrometer is said to be double focusing. In practice, manually adjusted *quadrupole lenses* (focus $x$, $F_x$, and focus $y$, $F_y$) at the entrance and exit faces are used to trim any deviation from ideal correction. One of the most important parameters of the spectrometer is the *dispersion*, $D$, which is the displacement of the focus per unit change in energy of the electron, as shown in *Figure 3.1*. It is typically 1–2 μm eV$^{-1}$. If the object is displaced normal to the beam direction, then to first order so is the image. The ratio of the displacements is the magnification and this will usually differ for displacements normal to, and in the symmetry plane of the spectrometer.

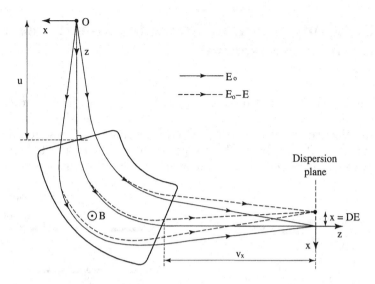

**Figure 3.1**. Schematic diagram showing the first-order focusing properties and dispersion for a 90° magnetic sector spectrometer.

Since the spectrometer is not rotationally symmetrical, its second-order aberration coefficients are not necessarily zero. However, by suitably curving the edges of the pole pieces the aberration coefficients can be changed and some of them made zero. Depending on the application, certain aberration coefficients are more important than others. Any residual effects may be corrected by *sextupole lenses* ($S_x$ and $S_y$) at the entrance and exit faces. Such a spectrometer is said to be second-order aberration corrected.

The energy resolution of the resultant EEL spectrum is determined by the energy spread of the electron source, the resolution of the spectrometer itself, the effect of external interference and the resolution of the detection system. For monochromatic electrons, the spectrometer will form an image in the dispersion plane. The width of this image along the dispersion direction divided by the dispersion gives the spectrometer resolution. To minimize aberrations, the entrance angle of the spectrometer must be limited – this is achieved by use of an adjustable *spectrometer entrance aperture* (SEA) at the position where the spectrometer bolts on to the camera chamber or, alternatively, by the use of post specimen apertures in the microscope to limit the collection angle of scattered electrons. The electron beam path must be also adequately screened from oscillating magnetic fields from both the mains supply and from high-frequency oscillations in the microscope; besides enclosing the camera chamber and spectrometer in a soft magnetic material such as mu-metal, this may be further achieved by adding an additional a.c. signal into the spectrometer excitation coils – usually termed *a.c. compensation correction*.

## 3.2 Coupling a magnetic sector spectrometer to the microscope

The simplest case is that of a dedicated STEM (or a TEM operated in STEM mode) with the specimen in field free space, as shown in *Figure 3.2a*. The electron probe is focused onto the specimen with a semi-angle of convergence, $\alpha$ (typically in the range 2–15 mrad). In most dedicated STEMs, there are no lenses after the specimen and the specimen is in the

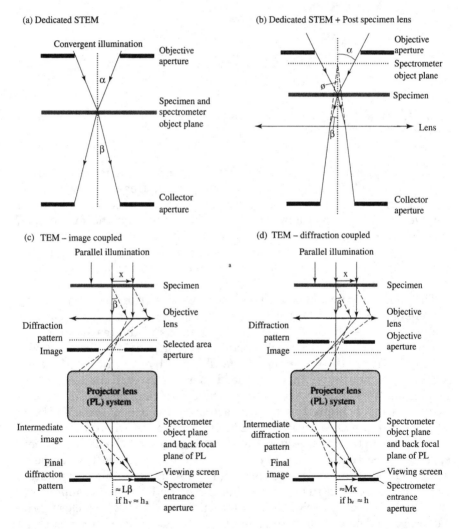

**Figure 3.2.** Schematic ray diagram of coupling between a magnetic prism spectrometer and: (a) a dedicated STEM; (b) a dedicated STEM with a post specimen lens; (c) a TEM – image coupled, and (d) a TEM – diffraction coupled. The diagram indicates the parameters that determine the area of analysis and the collection angle – see text for details.

object plane of the spectrometer. The effects of chromatic aberration are negligible thus the spatial resolution is independent of energy loss. The area irradiated by the probe becomes the spectrometer object and its size is determined by the size of the electron probe, which can be as low as 0.1 nm in recent instrumentation, and this effectively determines the spatial resolution. The spectrometer has a circular entrance aperture subtending a semi-angle, $\beta$, at the specimen (alternatively, in a dedicated STEM, a post-specimen collector aperture may be employed) and so the spectrometer entrance angle, $\phi$, is equal to $\beta$. Note that in a TEM/STEM operated in STEM mode, $\beta$ will be determined by the spectrometer entrance aperture and the camera length, which is discussed below.

For the STEM case outlined above, since the spectrometer entrance aperture is in the far-field, there is a diffraction pattern in its plane. If the magnetic field due to the STEM objective lens is strong and immerses the specimen, there will be magnetic induction after the specimen and this will act as a lens, giving rise to the situation in *Figure 3.2b*. Here the virtual image of the specimen is the spectrometer object and the angle of acceptance at the specimen, $\beta$, is $M\phi$ where $M$ is the magnification of the virtual image. Therefore, the angular range (collection angle), $\beta$, of electrons leaving the specimen can be compressed into a spectrometer of smaller angular acceptance, $\phi$, by forming a suitably magnified image of the specimen. However, while decreasing spectrometer aberrations and possibly increasing the energy resolution, this may increase overall aberrations due to those inherent in the post-specimen magnification system, as well as causing changes in the position of the spectrometer object plane and the source size for the spectrometer.

By using additional lenses, it is possible to maintain a fixed object plane for the spectrometer whilst allowing the angular compression to be varied over a wide range – this is similar to a conventional TEM operating in diffraction mode, shown in *Figure 3.2c*. Here the spectrometer entrance aperture is close to the microscope viewing screen on which the diffraction pattern is displayed. The spectrometer object is then the demagnified image of the specimen in the final projector lens and so the spectrometer is said to be *image coupled* (even though, when lowered, the TEM viewing screen would display a diffraction pattern). The angle of acceptance at the specimen, $\beta$, is defined by the radius of the spectrometer entrance aperture, $R$ (usually between 0.6 and 5 mm), so that, for inelastically scattered electrons,

$$\beta = (h_V R)/(h_A L) \tag{3.1}$$

where $L$ is the camera length (for the diffraction pattern displayed on the viewing screen) and $h_V$ and $h_A$ are the distances from the projector lens crossover to the viewing screen and spectrometer entrance aperture respectively. The area from which the spectrum is taken is limited either by the area of the illuminated region or by the size of the selected area diffraction aperture. However, both spherical and, more importantly, chromatic aberrations in both the TEM objective and projector lenses can affect the precision of the area selected using the SAED aperture and, for high spatial resolution analysis, limiting the area illuminated using a highly convergent probe (e.g. TEM microprobe or nanoprobe diffraction mode) is generally the preferred option.

The alternative TEM-based mode is *diffraction coupling* as illustrated in *Figure 3.2d*. This is the experimental arrangement used in EFTEM with a post-column imaging filter as discussed in Chapter 7. Here the microscope is in image mode and the spectrometer object is then the diffraction pattern at the crossover of the final projector lens. The diameter of the analysis area selected on the specimen, $D$, is usually controlled by the part of the image selected by the spectrometer entrance aperture. For a given entrance aperture, $R$, this area depends inversely on the image magnification, $M$, and is given by

$$D=(2Rh_v)/(Mh_A) \tag{3.2}$$

However, if the irradiated area in the image is smaller than that selected by the spectrometer entrance aperture, this will ultimately control the area of analysis but may cause problems in relative elemental quantification due to chromatic aberrations affecting the collection efficiency at different energy losses. The collection semi-angle can normally only be defined accurately by inserting the objective aperture – this can be calibrated for the range of objective apertures in a TEM using diffraction patterns from, say, a thin, polycrystalline gold film. However, use of the objective aperture may present problems when performing simultaneous EELS and EDX measurements owing to the stray X-ray signal generated from the aperture. Because of the effects of both spherical and chromatic aberrations of the post-specimen imaging lenses, recording an EELS spectrum without an objective aperture may cause problems in terms of both spatial and energy resolution.

The exact operating mode of the microscope and hence coupling to the spectrometer will depend on the nature of the problem being investigated; however, some general guidelines are discussed in Section 3.5. Further details are provided in Egerton (1996) and Disko *et al.* (1992).

# 3.3 Spectral recording

Having obtained a spectrum in the dispersion plane of the spectrometer, it is necessary to record the intensity at each particular energy loss as a sequence of digital values over a large number of detector channels (typically 1024 or 2048). Essentially the choice is between a single element detector, where the spectrum is sequentially scanned across the detector and recorded in a *serial* fashion, and a multi-element detector where the whole spectrum is recorded in *parallel* (Egerton, 1996; Disko *et al.*, 1992).

## 3.3.1 Serial recording

This is the original form of commercially available detection systems and is to be found on older EELS systems. Here the spectrum is scanned across a narrow slit in front of a scintillator–photomultiplier. Only those electrons passing through the slit reach the detector. The narrow slit is constructed from two pieces of high atomic number metal, which stop X-rays generated by the electrons from reaching the detector. The slit width is adjusted so that it matches the energy resolution of the spectrometer. If it is too narrow, signal is lost, while if it is too wide the energy resolution is degraded.

Four techniques are used to scan the spectrum:

(i)   Scanning the electrostatic potential on the drift tube in the spectrometer. This allows those electrons reaching the detector to follow the same trajectory. Furthermore it provides an independent method of energy calibration. This is the most common method employed.

(ii)  Scanning the magnetic field of the spectrometer by changing the current through the auxiliary coils wound within the main coils.

(iii) Ramping the accelerating voltage of the microscope. Electrons reaching the detector then travel along the same ray path so removing chromatic aberration effects. However, changing the incident energy causes defocusing which may change the illuminated area and intensity.

(iv)  Deflecting the spectrum at the exit of the magnet using suitable coils.

The most common form of serial detector is the scintillator–photomultiplier combination. The scintillator material converts electrons to visible photons, while the photomultiplier reconverts some of these photons into an output current. This arrangement is prefered to the direct exposure of, say, a solid-state silicon diode detector to energy loss electrons since radiation damage of the diode can occur at high electron doses. Cerium-doped YAG is the most commonly used scintillator material due to its radiation resistance. Overall, each incident electron yields about 20 photoelectrons simultaneously, compared to the single photolectron corresponding to a dark current event. At low signal levels, a pulse height discriminator can reject virtually all of the dark current events whilst registering nearly all of the real events.

The dynamic range of the signal level in an EELS spectrum is of the order of $10^8$ and, while the weakest signal can be pulse counted with dark current discrimination, the strongest signal in the low loss region will cause pulses to overlap even with the fastest electronics. Thus, in the low loss region, an analogue to digital converter (ADC) must be used. There is no ideal point for the spectrum to switch from one counting technique to the other and problems may arise trying to merge the two regions of the spectrum together. Hence, the switchover or *gain change* (see *Figure 2.1*) is normally done well away from any features of interest in the spectrum.

The EELS system is normally computer controlled so that the computer scans the spectrum in a series of equal steps, holds the spectrum at each position for a given dwell time and measures the intensity at each step. The choice of energy step for the spectrum depends on the energy resolution required. It is sensible to allot several channels per resolution element to observe detailed features. In the low loss region, the photomultiplier voltage is set to a pre-determined level and the computer records the output of the ADC; the latter is frequently a voltage to frequency converter whose output pulses are counted. At a point in the spectrum determined by the operator, the photomultiplier voltage is switched to a second pre-determined level and signal pulses that exceed the discriminator threshold are counted during the specified dwell time. Most systems allow a wide range of dwell times to be selected. Thus the operator can use a short dwell time and add an appropriate number of scans or use a long dwell time and take a single scan.

### 3.3.2 Parallel recording

While highly successful, serial recording is very inefficient despite having a detector quantum efficiency (DQE – effectively the noise performance of the detector) of between 0.5 and 0.9 for each channel; a perfect detector would have a DQE of 1. However, in a 1024-channel spectrum, as each channel is being recorded, the electrons in the remaining 1023 channels are being excluded giving an overall DQE of around 0.01 or less. Consequently, in recent years, parallel detectors have been developed which record the signal in all channels simultaneously. As a result, the technique has subsequently been renamed as PEELS.

Modern one-dimensional linear photodiode arrays (PDAs) and two-dimensional arrays of CCDs, the latter identical to those used for the digital recording of TEM images, are both well suited to the task of parallel recording. These devices are in the form of a large array of self-scanning silicon diodes which, once initially charged, can be discharged by the electron–hole pairs created either directly by the incident electrons or, indirectly, by photons created in a suitable scintillator. The amount of diode discharge depends on the local irradiation level during the period of exposure and, at the end of this integration period (which varies from typically 0.02 s (or below) to tens of seconds depending on the signal level), the diodes are interrogated and the amount of discharge on each diode is sequentially read out. This signal is superimposed on that due to thermal leakage currents as well as inherent electronic noise (together these are known as *dark current*) from each individual diode. The detector dark current contribution needs to be subtracted from the measured signal by integration of the electronic noise (i.e. with the beam excluded from the detection system) over an identical time period to that used for the signal measurement. Experimentally, the thermal leakage current is usually minimized by cooling the diode array to $-20°C$, however, this may result in incomplete diode read-out (a memory effect – particularly associated with intense signals such as ZLP) which may require the acquiring and discarding of several dark current read-outs prior to spectrum acquisition.

Direct electron counting on such arrays can cause problems due to the high yield of electron–hole pairs and radiation damage; thus, conversion of the incident electrons to photons (using scintillators as discussed for serial recording systems) and detection of the photons is currently the more popular approach for PEELS detection. *Figure 3.3* shows a schematic detector assembly using a fibre-optic plate to couple the scintillator to the diode array. Such a detector has a high DQE for a signal with a dynamic range of $10^4$ provided that the highest signal is close to the diode output saturation level; CCDs have slightly higher dynamic ranges due to their two-dimensional nature. The integration time for acquiring and reading out the signal is therefore chosen to fulfil this criterion – for a typical one-dimensional photodiode array in the Gatan PEELS system this saturation level is around 16 000 counts. Thus if all the signal of interest possesses a suitable dynamic range, it can be recorded simultaneously. However, owing to the large dynamic range, the whole EEL spectrum (0–2 keV) would generally require at least two or three suitable integration periods. One major problem with indirect exposure systems is the trapping of charge in the scintillator following intense exposure, which can temporarily increase the electron–photon conversion efficiency and cause 'ghost or memory' features similar to those arising from diode read-out – these can be removed by prolonged exposure of the scintillator to an undispersed beam at levels close to saturation (known as 'flooding' the detector).

The detection elements in such a multi-element detector are adjacent and the spreading that occurs when high-energy electrons stop in a solid material limits the spatial resolution of such a detector (see Section

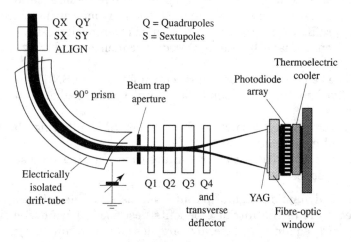

**Figure 3.3.** Schematic diagram of a Gatan PEELS system. Redrawn from Egerton (1996) *Electron Energy Loss Spectroscopy in the Electron Microscope,* 2nd Edn, with permission of Plenum Publishers.

1.2.6). This lateral spreading is of the order of 30 μm for high-energy electrons so that the dispersion must be increased to a value greater than $30/\Delta E$ μm/eV where $\Delta E$ is the desired energy resolution. The inclusion of a set of quadrupole lenses after the spectrometer allows the dispersion plane of the spectrometer to be electron optically magnified. The non-isotropic focusing properties of these lenses allow the spectrum to be in focus in the dispersion direction, but highly defocused in the perpendicular direction. Thus the magnification in the dispersion direction can be varied but the width in the perpendicular direction can be held relatively constant. Hence the current density on the scintillator is reduced and minor blemishes averaged. The same area of the detector is always used, easing the problem of correction of channel to channel variations in sensitivity (gain) that are always present in such arrays and lead to so-called fixed pattern noise. However, for modern detector arrays this variation is usually less than 1% and may be corrected by uniform exposure of the array or by scanning or shifting the spectrum across the array, realigning the spectra and averaging (Boothroyd *et al.*, 1990). A further correction necessary in high-energy resolution systems is due to the apparent degradation in resolution caused by the spreading of electrons and light in the scintillator. The system point spread function (PSF) can be measured by reducing the dispersion to such a value that all the elastically scattered electrons falling within the ZLP are focused to a spot smaller than the physical dimension of a single diode. A more reliable method employs the modulation induced on a sharp edge. The tails on the measured PSF function arise from the spreading of the beam by the scintillator.

In comparison with serial recording, parallel detection increases the detection efficiency in EELS by a factor of about 100. The benefits of such an enormous increase in efficiency may be summarized as follows:

(i)   Reduced recording time for a given signal-to-noise ratio (SNR) allowing more points to be obtained when say measuring concentration profiles, or resulting in less contamination, drift and/or radiation damage to the analysed specimen area during a given data acquisition.

(ii)   Reduced electron dose on the specimen for a given SNR allowing an improved spatial resolution to be obtained with a radiation sensitive material.

(iii)   Improved energy resolution for a given SNR allowing the more detailed measurement of ELNES.

(iv)   Improved SNR for a given recording time allowing the use of less intense edges.

The reduced dynamic range of parallel versus serial detection systems may offset some of the intrinsic benefits described above owing to the need to record sequentially several energy loss regions. However, parallel detection does provide a true average of the spectrum over the recording period and so avoids any distortion due to changes in conditions during serial recording.

EELS data from the one- or two-dimensional detector array is collected and stored on a multi-channel analyser-type computer software package. The data can be processed using appropriate computer software such as the commercially available Gatan EL/P package.

## 3.4 Energy-filtered imaging

Energy-filtered TEM, as well as alternative scanning-based techniques, are discussed in more detail in Chapter 7. Here a brief description of the most common experimental arrangements is provided for completeness, with further information being found in Reimer (1995).

The design of the post-column imaging filter manufactured by Gatan (as opposed to the in-column Castaing–Henry and $\Omega$ filters) is again based on a magnetic sector spectrometer and is shown in *Figure 3.4*. The final projector lens crossover is in the spectrometer object plane; however, instead of having a detection system at the first spectrometer image plane (where the spectrum is formed), here a variable energy selecting slit (between say 1 and 100 eV in width) is inserted. This results in the formation of an energy selected image (or, alternatively, a diffraction pattern – if the microscope is operated in diffraction mode) at the second image plane of the spectrometer. Unfortunately, the spectrometer is only double-focusing at the spectrum plane and the image needs to be corrected for aberrations and astigmatism via the use of post-spectrometer quadrupoles and sextupoles shown in *Figure 3.4*. Suitable excitation of the quadrupoles can be used to project the energy-filtered image or diffraction pattern onto either a TV rate or slow-scan CCD camera (imaging mode), alternatively the slit may be retracted and the

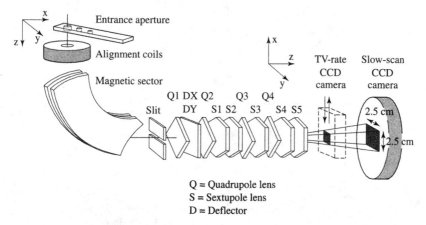

**Figure 3.4.** Schematic diagram of a Gatan imaging filter. Redrawn from Egerton (1996) *Electron Energy Loss Spectroscopy in the Electron Microscope*, 2nd Edn, with permission of Plenum Publishers.

spectrum plane may be projected onto the CCD to record or focus the spectrum (spectroscopy mode).

The current commercial in-column imaging filter manufactured by Zeiss (LEO) is generally based on the $\Omega$ filter design (and variations thereon) which involves the use of four magnetic prisms arranged symmetrically in the form of a Greek letter $\Omega$. As the filter is located in the middle of the TEM column, between the objective lens and the main projector lens system as shown in *Figure 3.5*, it forms an integral part of the whole TEM design and is not simply an add-on attachment like the post-column GIF described in the last paragraph. The $\Omega$ design allows the beam to emerge along the original optic axis and the inherent symmetry about the mid-plane of the prism system causes various second order

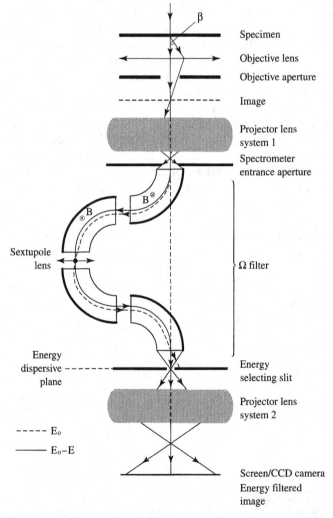

**Figure 3.5.** Schematic diagram of an $\Omega$ filter in energy-filtered imaging mode.

aberrations to vanish and the remaining can, in principle, be compensated for via use of sextupoles positioned symmetrically in the mid-plane. With the microscope in image mode, the object point of the filter system contains a diffraction pattern, which forms an EEL spectrum in a conjugate plane below the filter. Insertion of an energy selecting slit in this energy-dispersive plane and subsequent projection of the corresponding image plane onto a screen or CCD camera allows energy-filtered imaging. Alternatively focussing the projector lens system on the energy-dispersive plane permits EEL spectroscopy. Finally operation of the microscope in diffraction mode and projection of the filter image plane, allows energy-filtered diffraction patterns to be recorded.

The operating performance of both corrected post-column and in-column energy filters are generally similar; however the added flexibility of the add-on nature of the former design has led to their more widespread use over a range of different operating voltages. The operation of both filter designs attempts to minimize the effects of chromatic aberrations in the post-filter lenses, this involves selection of the desired energy loss range by changing the microscope accelerating voltage rather than altering the spectrometer drift tube voltage.

Finally a single magnetic prism spectrometer and detector system attached to a STEM can also be used to form energy-filtered images (Hunt and Williams, 1991). Either a complete EEL spectrum can be recorded at each position of the scanned probe and the image formed via post-acquisition spectrum processing (known as spectrum imaging), or more simply the output of a selected number of detector channels can be directed to an image-display monitor which is synchronized with the probe scan coils.

# 3.5 Choice of experimental conditions for EELS

## 3.5.1 Experimental parameters in the TEM/STEM

The parameters in the TEM have been briefly alluded to in Section 1.3. Here we deal with those directly relevant to EELS. Briefly these may be conveniently divided as follows.

Those concerned with the electron source:
(i)   Choice of source (e.g. W, $LaB_6$, thermally assisted or cold field emission) and emission current settings – these affect the energy spread of the electron beam, the overall intensity and the brightness of the incident electron beam.
(ii)  The incident beam energy, $E_o$. Higher accelerating voltages and hence beam energies generally give increased brightness from the electron gun and allow the study of thicker samples due to the increased inelastic mean free path (see Equation 4.2). In terms of

radiation damage of the specimen, increased knock-on damage may occur at higher voltages, although ionization-based damage mechanisms will decrease with increasing voltage. Another advantage of lower beam energies is the increased spectral resolution due to the improved performance of the spectrometer/detector system and the reduced energy spread of the electron beam.

Those concerned with the mode of operation:

(i)   Use of parallel or focussed illumination via excitation of condenser lenses – this governs the convergence semi-angle, $\alpha$ and the analysis area. The value of $\alpha$ can be determined for a given condenser lens current by measuring the angular width of diffraction discs from a material with a known $2\theta_{Bragg}$ (e.g. single crystal $<111>$ oriented gold film), where $\theta_{Bragg}$ is the Bragg angle for diffraction from a set of planes of spacing, $d$, using electrons of wavelength $\lambda$ and $\sin \theta_{Bragg} = \lambda/2d$.

(ii)  Operation of a CTEM in either imaging or diffraction mode and choice of magnification or camera length – generally for routine EEL spectroscopy from relatively large areas either diffraction mode with an SAED aperture or low-magnification image mode gives higher signals. Diffraction mode is the most suitable for EELS measurements due to the reduced problems associated with chromatic aberration in the objective and projector lenses which can result in differing collection efficiencies at different energy losses. Chromatic aberrations may be reduced by altering the microscope accelerating voltage so as to shift the spectrum across the detector rather than changing the drift tube voltage. For high spatial resolution EEL spectroscopy, operation in diffraction mode with a convergent beam is recommended if adequate signals from well-defined specimen areas are to be achieved (i.e. microprobe or nanoprobe diffraction modes). When operating in CTEM diffraction mode, care must be taken to centre the diffraction pattern about the spectrometer entrance aperture (SEA) which will be close to the centre of the microscope viewing screen. Generally, in both imaging and diffraction modes, this involves using the microscope beam/diffraction shift controls to maximize the measured EELS signal. Use of a two-dimensional detector with post-spectrometer lenses, such as a GIF, allows the direct viewing of both the image or diffraction pattern and the SEA on the camera.

(iii) Use of scanning mode. In a dedicated STEM, the STEM bright field detector is on-axis and must be retracted before electrons can enter the on-axis EEL spectrometer; however, simultaneous EELS and dark field (DF) imaging using the ADF detector is possible. In a TEM/STEM hybrid instrument the BF/ADF STEM detector is usually off-axis and the scanned beam needs to be deflected for STEM imaging; here simultaneous STEM imaging and on-axis EELS is not possible without the use of on-axis ADF STEM detector, usually positioned above the TEM viewing screen.

Those concerned with use of apertures:

(i) Choice of condenser apertures – this affects overall beam intensity, convergence angle and degree of coherency. Such apertures can also collimate the beam and remove stray scattering from the electron gun assembly which may possess different incident energies and contribute artefacts to the EEL spectrum.

(ii) Use of an objective aperture – with the microscope in imaging mode the choice of objective aperture governs the value of the collection angle. The angular range of a set of TEM objective apertures can be calibrated by simultaneously viewing a known diffraction pattern (e.g. polycrystalline gold) and the outline of the objective aperture. The aperture semi-angle, $\beta$, can be referenced to the known $2\theta_{Bragg}$ angles of the radii of the ring pattern.

(iii) Use of selected area diffraction aperture – in CTEM diffraction mode, this defines the analysis area provided the beam is not highly convergent. In the latter case (e.g. microprobe or nanoprobe diffraction mode), an SAED aperture can reduce stray gun scattering.

(iv) Choice of spectrometer entrance aperture – with the microscope in CTEM diffraction mode this, together with the camera length, governs the collection angle, $\beta$. In CTEM image mode with a relatively parallel beam, the SEA and the microscope magnification govern the analysis area.

## 3.5.2 *Experimental parameters in EELS*

The main experimental parameters in EELS are:

(i) Energy resolution.
(ii) SNR.
(iii) Signal-to-background ratio (SBR).

In principle, optimal data would have the highest possible energy resolution and also maximize both the SNR and the SBR to give the necessary statistical accuracy and information content; however, it is important to realize that these individual factors are not independent.

Generally the spectral energy resolution is defined as the FWHM of the ZLP; however, it is important to realize that the energy resolution at a particular energy loss feature may well be degraded from that obtained at the ZLP owing to the energy-dependent focusing properties of the spectrometer. As already discussed, the overall energy resolution is mainly a function of the type of electron source, filament emission current, the spectrometer resolution, external fields and the energy dispersion at the detector. Assuming external fields have been minimized and a sufficiently high energy dispersion is employed, the spectral energy resolution is primarily determined by the type of emitter – the energy spread will be of the order of 1.5–3 eV for a tungsten emitter, 1–1.5 eV for a $LaB_6$ emitter, 0.6–0.8 eV for a thermally assisted field emitter and approximately 0.3 eV for a cold field emitter. Reducing the emission current, that is, desaturating a thermionic emitter and also reducing the extraction voltage on a Schottky emitter, reduces the energy spread and hence improves spectral resolution at the expense of the SNR.

It is becoming increasingly common to deconvolve EELS data with an energy resolution function (usually a zero loss (ZL) spectrum measured with the beam passing through a hole in the specimen) so as to improve the apparent energy resolution, although, as with all deconvolution techniques this is usually at the expense of some added noise. Some specialized microscopes employ electron monochromators to improve energy; however, this can dramatically cut down the total electron beam intensity. As discussed in Section 3.1, the spectrometer resolution improves with decreasing collection angle, β. Generally for a given electron source, increasing the energy resolution will decrease the SNR.

For an efficient PEELS detection system, the spectral noise is principally determined by the square root of the number of electrons detected – the so-called shot noise. The fractional noise or SNR is therefore dependent on the number of electrons arriving at the detector and will be a function of the electron intensity incident at the specimen, and hence also the incident beam energy, the brightness of the source, the emission current, the size of condenser aperture and the collection angle of electrons entering the spectrometer (which may be dependent on either the size of objective aperture or the camera length/spectrometer aperture depending on the microscope operating mode). In PEELS the SNR also depends on the optimum use of the dynamic range of the detection system, which involves the choice of a suitable integration time before data read-out so that the signal is just below the saturation level of the detector. For inner-shell ionization, the SNR generally increases with increasing beam energy as does, in most cases, specimen damage (see Section 3.6) – an optimum regime for EELS of inorganic specimens lies in the range 100–300 keV, much higher than the case for organic specimens.

A high value of the SBR improves the visibility of EELS edges on the steep background and improves the reliability of background subtraction procedures and detection sensitivity. The SBR is a function of specimen thickness owing to the large plasmon contribution to the background, as well as the collection angle, β. This is discussed in Chapters 4 and 5. Using a small collection angle increases the SBR, since a large proportion of the background intensity under a core loss edge originates from excitation of valence electrons. At high energy loss, the angular distribution of these low-energy excitations is peaked at high angles (this is the Bethe ridge – see Section 2.4) and the background contribution to an edge is therefore maximized at large scattering angles. However, there is a trade off between SBR and SNR since reducing β will reduce the SNR. Typically collection angles of around two or three times $\theta_E$ are employed as a compromise, leading to values of the order of 10 mrad for most commonly observed energy loss features in the high loss regime. This choice of collection angle has the added advantage of ensuring that electron transitions lie within the dipole-allowed regime, which has importance for both elemental quantification and studies of ELNES at core loss ionization edges described in Chapters 5 and 6. Smaller collection angles not only increase noise but may also isolate anisotropic effects (see Section 8.1) which may be undesirable, depending on the exact analytical application.

The accurate determination of the absolute energy loss of EELS features is also an important consideration, both for simple edge identification and for the observation of valence-dependent chemical shift effects (see Section 6.2). Exact energy calibration can be difficult, particularly for PEELS detection systems where it is usually only possible to record portions, rather than the whole EELS spectrum. The relative energy-loss scale in an EELS spectrum depends critically on the accurate calibration of the spectrometer dispersion settings, and this is usually achieved by applying known voltages to the spectrometer drift tube and observing the displacement of a particular feature of known energy loss, say the ZLP (obviously at 0 eV) or the Ni $L_3$-edge in an NiO thin film (at 853.2 eV), across the detector array. Alternatively, in current EFTEM systems, it is also often possible to vary the microscope high voltage while keeping the drift tube voltage constant and again record the displacement of spectral features. Then, in order to determine the absolute energy of an EELS feature of unknown energy, the drift tube voltage (or change in microscope high voltage) required to displace an EELS feature to the prior position of the ZLP on the detector is measured. This results in an accuracy in the absolute energy of at best $\pm 1$ channel (i.e. 0.1–0.3 eV depending on the exact spectrometer dispersion setting). It is important to realize that significant energy drift of the spectrum across the detector can occur during measurements over extended periods of time, making absolute energy determination unreliable. This energy drift arises principally as a result of hysteresis effects or changes in the external magnetic field (due to, for example, movement of the operator's metal chair – wooden chairs are therefore recommended!). Spectral drift may also occur as a result of fluctuations in the high-voltage or spectrometer power supplies and may be overcome by the use of feedback circuits (Bleloch *et al.*, 1999); this approach has the added advantage of improving energy resolution. Where possible, one experimental method of circumventing these problems is to use a suitable internal calibration in the same spectrum such as the EELS signals arising from carbon contamination (the leading $\pi^*$ peak maximum on the C *K*-edge in graphite lies at 285.4 eV), oxide layers or another feature of known energy loss.

The experimental parameters described above have relevance for not only EELS spectroscopy but also for energy-filtered TEM discussed in Chapter 7. In EFTEM a further factor which influences the SNR in energy-filtered images is the size and position of the energy-selecting window. Berger and Kohl (1993) have published some recommendations for the choice of experimental conditions in EFTEM elemental imaging.

# 3.6 Specimen parameters

Owing to the problems associated with multiple inelastic scattering, the technique of EELS depends critically on the preparation of specimens with sufficiently thin areas for analysis. Very roughly, useful spectral data may be obtained from areas with a projected thickness of less than one inelastic mean free path, $\Lambda$ (of the order of 100 nm at normal beam energies) as discussed in Section 4.1, although the use of deconvolution techniques are advisable for data from all but the thinnest regions, say $0.3\Lambda$ and above. The use of standard TEM preparation techniques are often sufficient to meet these needs, provided the material microstructure is unaffected by the thinning process. Focused ion beam (FIB) preparation techniques can provide large sample regions of very uniform thickness. The use of higher accelerating voltages allows analysis of thicker specimens via increases in $\Lambda$, although this is not a linear relationship.

Specimen orientation can also be an important consideration, as orientations which give rise to strongly diffracting conditions (e.g. the appearance of intense Bragg diffraction spots) can affect the accuracy of elemental quantification (Chapter 5), change ELNES intensities (Chapter 6), distort EFTEM maps (Chapter 7) and increase beam broadening in high spatial resolution analysis (Chapter 8). A usual procedure is to tilt the sample just off a major zone axis.

Contamination, usually in the form of hydrocarbon deposition, migration and polymerization during analysis, may also be a serious problem. This rapidly makes the specimen too thick to identify other edges and causes changes in the spectral background during acquisition. Great care must be taken to ensure that the microscope and specimen holder are free of hydrocarbons. Specimen preparation and handling should always seek to exclude sources of contamination, such as industrial quality solvents. A 'cold finger' or cooled specimen stage held at liquid nitrogen temperature can help minimize this problem, or, alternatively, various pre-treatments of the sample, such as gentle baking or plasma cleaning, may be employed with direct transfer under vacuum into the microscope column. Alternatively, the whole specimen can be flooded with a large dose of electrons so as to fix the hydrocarbon in place as a very thin, uniform layer prior to the use of focused probe techniques for data acquisition.

The ultimate limitation in EELS experiments arises from damage to the specimen induced by the high-energy incident electron beam. This is dependent on many factors (Williams and Carter, 1997) including: the nature of the specimen, the electron dose per unit area (which can be extremely large if high spatial resolution is required), the incident beam energy and the acquisition time required for adequate counting statistics. In addition to direct heating of the sample, crystalline samples may lose their structural order and light atoms may be knocked out of their lattice sites and lost from the sample area under study. Depending on the type of material and the particular damage mechanism involved, damage rates may be either reduced or increased by increasing the accelerating voltage (see Section 3.5.1) – although at sufficiently high voltages, typically above 400 keV, all materials will damage to some extent. At sufficiently high current densities the electron beam will drill holes in some specimens, the rate of mass loss varying with the diameter of the irradiated volume, the specimen thickness and the incident current. Rapid acquisition parallel recording of spectra may be employed to form *time-resolved spectral series* and these may be used to study the nature of the damage processes occurring.

## 3.7 Summary of experimental set-up and data acquisition procedures for EELS

This very basic list is primarily concerned with PEELS, although serial systems follow the same basic steps. More detailed considerations for EFTEM imaging are given in Chapter 7.

(i)　Align microscope at chosen accelerating voltage.

(ii)　With no specimen present – find ZLP, focus and align ZLP (e.g. adjust $x$ and $y$ quadrupoles and sextupoles).

(iii)　Correct a.c. compensation.

(iv)　For EFTEM, tune imaging filter optics and align energy selecting slit.

(v)　Select and, if necessary, calibrate spectrometer dispersion manually or using an automated software routine.

(vi)　Calibrate detector gain response if required (see below).

(vii)　Insert specimen and select mode of operation (CTEM diffraction or imaging modes, or STEM).

(viii)　Select analytical area using scanned area (STEM), SAED aperture and/or condenser excitation (CTEM diffraction mode), or SEA and microscope magnification (CTEM image mode).

(ix)　Select orientation on specimen (generally avoid strongly diffracting conditions).

(x)　Select convergence angle using condenser aperture/condenser excitation.

(xi)　Select collection angle using either objective aperture (TEM image mode), camera length and SEA (TEM diffraction mode and STEM).

(xii)　Centre image or diffraction pattern about SEA.

(xiii)　Select drift tube/high voltage offset.

(xiv)　Record data using suitable integration time (close to detector saturation level) and then measure corresponding dark current with beam blanked.

(xv)　Determine absolute energy loss using drift tube/high-voltage offset or internal energy calibration.

# 3.8 Summary of data correction procedures for EELS

PEELS data requires a number of pre-processing steps prior to accurate analysis and quantification:

(i)   Correction of raw data for detector artefacts (e.g. dark current) and detector response characteristics (e.g. channel to channel gain variations), as well as the point spread function of the detector system.

(ii)  Correction of the data for the effects of excessive sample thickness, this involves deconvolution of the intensity component which has undergone both plasmon and ionization events (termed the multiple inelastic scattering contribution). The deconvolution techniques employed to produce a single scattering distribution are described in Chapter 4.

(iii) Deconvolve energy resolution function (i.e. experimental ZLP) if required.

Similar correction procedures for EFTEM images are summarized in Chapter 7.

# References and further reading

**Berger, A. and Kohl, H.** (1993) Optimum imaging parameters for elemental mapping in an energy filtering TEM. *Optik* **92:** 175–193.

**Bleloch, A., Brown, L.M., Marsh, M.J., McMullen, D., Rickard, J.J. and Stolojan, V.** (1999) Cancelling energy drift in a PEELS spectrometer. *Inst. Phys. Conf. Ser.* **161:** 195–198.

**Boothroyd, C.B., Sato, K. and Yamada, K.** (1990) The detection of 0.5atom% boron in $Ni_3Al$ using parallel electron energy loss spectroscopy. *Proc XIIth Int. Cong. Electron Microscopy*, Vol. 2. San Francisco Press, San Francisco, pp. 80–81.

**Disko, M.M., Ahn, C.C. and Fultz, B.** (1992) *Transmission Electron Energy Loss Spectrometry in Materials Science*. TMS, Warrendale, Pennsylvania.

**Egerton, R.F.** (1996) *Electron Energy Loss Spectroscopy in the Electron Microscope*. Plenum Press, New York.

**Hunt, J.A. and Williams, D.B.** (1991) Electron energy loss spectrum imaging. *Ultramicroscopy* **38:** 47–73.

**Reimer, L. (ed.)** (1995) *Energy Filtering Transmission Electron Microscopy*. Springer, Heidelberg.

**Williams, D.B. and Carter, C.B.** (1997) *Transmission Electron Microscopy*. Plenum Press, New York.

# 4 Low loss spectroscopy

The low loss regime contains the majority of electrons in a typical EELS spectrum and therefore has the experimental advantage of possessing high signal. The low loss region contains much of the solid state character, that is, the response of the weakly bound, often collective valence electrons, of the solid to the disturbance created by the incident electron beam; in addition, ionization of electrons in low binding energy atomic-like levels also occurs in this energy loss range. Below we summarize a number of important aspects of this spectral region.

## 4.1 Quantification of sample thickness

As the sample thickness increases, the probability of inelastic scattering also increases, particularly plasmon excitation, which possesses the highest cross-section of all inelastic processes. The frequency of these independent scattering events follows a Poisson distribution and, from this behaviour it can be shown that the ratio of the total intensity in the spectrum, $I_T$, to the intensity in the elastic ZLP, $I_{ZL}$, gives a measure of the specimen thickness, $t$, via the formula

$$t = \Lambda.\ln(I_T/I_{ZL}) \qquad (4.1)$$

where $\Lambda$ is the mean free path for inelastic scattering (effectively the distance travelled by the fast electron between inelastic collisions, which is of the order of 100 nm at 100 keV). The two intensities shown in *Figure 4.1* may be determined via simple integration of the spectral data using an appropriate software package. The accuracy of thickness determination using EELS depends critically on the value of the mean free path employed. Plasmon mean free paths and total mean free paths have been tabulated for various materials and incident beam energies (via an independent determination of sample thickness using alternative methods such as convergent beam electron diffraction etc.), they may also be calculated approximately using the free electron formula,

$$\Lambda = a_0/[\gamma\theta_E \ln(\beta/\theta_E)] \qquad (4.2)$$

where $\gamma$ is the relativistic factor (Equation 1.1), $\beta$ is the experimental collection semi-angle and $\theta_E$ is the characteristic scattering angle for plasmon excitation. Various other formulae and parameterization schemes for calculating inelastic mean free paths are discussed in Egerton (1996).

With increasing sample thickness, there is an increased probability for (single) plasmon excitation as well as increase in the probability of multiple inelastic scattering, that is, a double or triple plasmon excitation, as well as an inner-shell ionization event followed by a plasmon excitation. These effects are shown schematically in *Figures 4.1* and *4.2*. It is possible to remove this multiple inelastic scattering contribution from either the whole EELS spectrum or a particular spectral region by Fourier transform *deconvolution* techniques. Two techniques are routinely employed, one, known as the *Fourier-Log method*, requires the whole spectrum over the whole dynamic range as

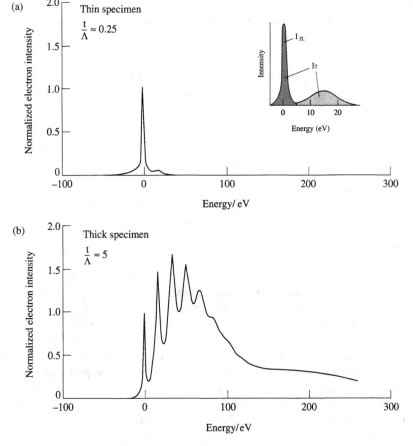

**Figure 4.1.** Schematic EELS spectra for silicon from regions of different thickness: (a) $t/\Lambda \approx 0.25$ and (b) $t/\Lambda \approx 5$. The relevant areas, $I_{ZL}$ and $I_T$, for thickness determination are shown inset in (a).

input data. This large signal dynamic range can be a problem with data recorded in parallel as is discussed in Section 3.3.2. The second, known as the *Fourier-Ratio method*, requires a spectrum containing the feature of interest (i.e an ionization edge) which has had the preceding spectral background removed (see Section 5.2). A second spectrum containing the low loss region from the same specimen area is then used to deconvolute the first spectrum. Both these techniques essentially provide as output a whole spectrum (Fourier-Log method) or spectral region (Fourier-Ratio method) corrected for the effects of thickness – known as a *single scattering distribution* (SSD), which can be important if accurate elemental quantification or detailed analysis of ELNES fine structure is to be achieved. The two mathematical schemes can be implemented as plug-in subroutines in data acquisition and processing packages and are discussed more fully in Egerton (1996). Deconvolution dramatically improves the signal to background ratios for the weak spectral features such as core loss edges, although this may be at the expense of some added noise. Examples of the effect of deconvolution are shown in *Figure 4.2*. As a rough rule of thumb, spectra recorded from regions where $t/\Lambda$ exceeds 0.3–0.5 (see Equation 4.1) generally require deconvolution of multiple inelastic scattering if any reliable quantitative analysis is to be achieved.

# 4.2 Quantitative aspects of low loss data

Generally the low loss region is dominated by the *bulk plasmon* excitation and this may be thought of as a resonant oscillation of the valence electron gas of the solid (as pictured in a Drude–Lorentz model) in response to the fast incident electron. Within the free-electron model of solids an expression for the bulk plasmon energy, $E_p$, is given by

$$E_p = h/2\pi \, [Ne^2/(m_e\varepsilon_o)]^{1/2} \tag{4.3}$$

where $N$ is the valence electron density and $\varepsilon_o$ the permittivity of the free space. This free electron formula works suprisingly well for a range of elements and compounds; a list of some experimental plasmon energies is provided in Egerton (1996).

For free-electron metals, such as aluminium, there is very little damping of the oscillation and the observed plasmon lines are very sharp and their position is dependent on the square root of the valence electron density. Such a spectrum is shown in *Figure 4.3a* for metallic Al, where $E_p$=15.3 eV. The successive lines in the plasmon spectrum correspond to multiple inelastic scattering (due to excessive sample thickness) and occur at harmonics of the plasmon energy. For accurate analysis, low loss data should always be deconvoluted for the effects of multiple inelastic scattering as described in Section 4.1. In insulators and semiconductors,

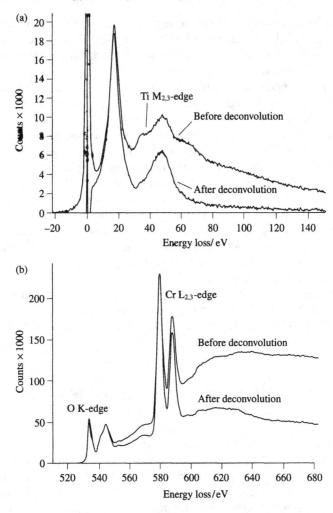

**Figure 4.2.** The effects of (a) Fourier-Log deconvolution on the low loss spectrum of a titanium alloy and (b) Fourier-Ratio deconvolution on the O *K*-edge in chromium oxide, $Cr_2O_3$. Note for the Fourier-Log deconvoluted data in part (a) the ZLP has been fitted and removed from the low loss spectrum and replaced with a delta function. In the case of the Fourier-Ratio deconvolution in part (b), the background preceding the O *K*-edge has been removed prior to deconvolution.

plasmon lines are considerably broader than in metals since the valence electrons are no longer free and this is seen in *Figure 4.3b* for the case of diamond. This may be analysed in terms of the free-electron model via inclusion of a relaxation time, $\tau = 1/\Gamma$ ($\Gamma$ is the damping constant), to account for the damping of the plasmon oscillation by scattering with the ion-core lattice. This results in a plasmon of width proportional to $1/\tau$.

As the plasmon energy and hence plasmon peak position is sensitive to the valence electron density, any changes in this quantity, such as those due to alloying in metals or general structural rearrangement in different microstructural phases, can be detected as a shift in the plasmon energy

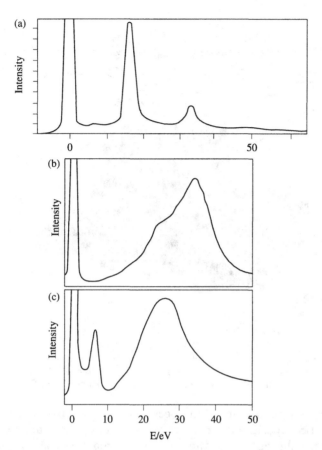

**Figure 4.3.** Low loss spectra of: (a) aluminium, note the presence of a surface (oxide) plasmon peak below 10 eV; (b) diamond, note the presence of a band gap at about 6 eV; and (c) graphite, note the presence of a $\pi-\pi^*$ interband transition at 6.2 eV. In all cases the ZLP extends off the plot.

allowing a means of *phase identification*. *Figure 4.4* shows a spatial map of the plasmon peak position in an Al–Li alloy, this is a processed spectrum image as described in Section 7.2 recorded on a STEM which reveals the distribution of lithium in a 400-nm$^2$ area (Disko *et al.*, 1992). Here, shifts in the plasmon peak position depend on the valence electron density, and hence the Li concentration (nominally Li contributes one electron per atom to the valence electron gas as opposed to three electrons per atom in the case of aluminium); the peaks correspond to precipitates of the distinct $\partial'$ Al$_3$Li phase. In a microstructure, changes in properties such as electrical or thermal conductivity and even elastic modulus are a function of the local valence electron densities, hence determination of the plasmon energy can provide a powerful tool for localized property determination in solid microstructures (Monthioux *et al.*, 1997).

In addition to three-dimensional bulk plasmons it is also possible to excite two-dimensional *surface plasmons* or *interface* plasmons; the

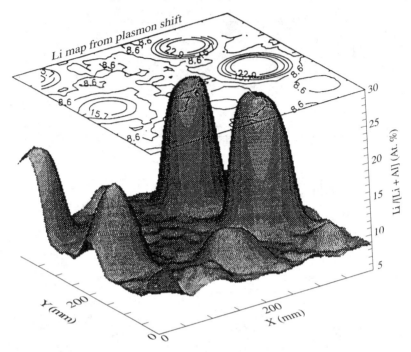

**Figure 4.4.** Contour and surface plot of lithium concentration in an Al-10.5at%Li alloy determined by shifts in the plasmon peak position. Redrawn from Disko *et al.* (1992) *Transmission Electron Energy Loss Spectrometry in Materials Science,* with permission of C. Ahn.

former are apparent in very thin specimens, small nanoparticles or in reflection EELS experiments. For the simplest case of a planar vacuum/metal interface the energy of the surface plasmon can be shown to be $1/\sqrt{2}$ times the bulk plasmon energy; there are other formulae applicable to more complicated geometries and materials combinations which may be found in Egerton (1996).

As well as plasmon oscillations, the low loss region may also exhibit *interband transitions*, that is, single electron transitions from the valence band to unoccupied states in the conduction band, which appear as peaks superimposed on the main plasmon peak. In a solid state physics picture, these interband transitions represent a joint density of states (JDOS) – a convolution between the valence and conduction band DOS. The presence of these single-electron excitations can lead to a shift in the energy of the plasmon resonance. The low loss spectrum of graphite in *Figure 4.3c* exhibits a π to π* interband transition at 6.2 eV in addition to the main plasmon peak at 27 eV. This interband transition causes a shift of the free electron plasmon energy to higher energies since these electrons also contribute to the resonant oscillation. If the energy of the interband transition is greater than that of the free electron plasmon energy this causes a corresponding shift of the plasmon peak to lower energies. In an insulator with a *band gap* there should no interband transitions below the band gap energy and thus the band gap is associated with an initial rise in intensity in the low loss region – seen in the low loss spectrum of diamond in *Figure 4.2b* at about 6 eV. Accurate removal of the ZLP from the rest of the spectrum allows the determination of band gap energies and whether the gap is direct or indirect (Rafferty and Brown, 1998); however, a more accurate procedure is based on extraction of the dielectric function of the material, described below (Daniels *et al.*, 1970; Raether, 1980).

A more sophisticated analysis of the low loss region is based upon the concept of the *dielectric function*, ε, of the material. This is a complex quantity, dependent on both energy loss and momentum, which represents the response of the entire solid to the disturbance created by the incident electron. The same response function, ε, describes the interaction of photons with a solid and this means that energy loss data may be correlated with the results of optical measurements in the visible and UV regions of the electromagnetic spectrum, including quantities such as refractive index, absorption and reflection coefficients.

This approach is related to the Bethe formulation, which was introduced in Section 2.4. To briefly recap, the exact form of the EELS spectrum is given by the double differential cross-section, $d^2\sigma/(dE\,d\Omega)$. Equation (2.4) reveals that this quantity may be decomposed into a term reflecting scattering from a free electron (the Rutherford cross-section) as well as a dynamic structure factor for the atom in a solid summarized by the generalized oscillator strength (GOS) per unit energy loss, $df/dE$. Ritchie (1957) showed that the Bethe and dielectric descriptions are equivalent if

$$df/dE = 2E/(\pi E_p^2).\mathrm{Im}[-1/\varepsilon] \qquad (4.4)$$

The last term, $\mathrm{Im}[-1/\varepsilon]$, is known as the *energy loss function* and provides a complete description of the response of the solid to the incident fast electron – essentially it describes the form of the low loss EELS spectrum (and, in principle, the higher loss region as well) corrected for the effects of thickness. For the free electron gas model and for small values of $q$, the energy loss function can be regarded as being independent of $q$ and is given by

$$\mathrm{Im}[-1/\varepsilon] = (E.\Gamma.E_p^2)/[(E^2 - E_p^2)^2 + (E.h/(2\pi).\Gamma)^2] \qquad (4.5)$$

where $\Gamma$ is the damping constant. The maximum in the energy loss function occurs at $[E_p^2 - (h\Gamma/4\pi)^2]^{1/2}$ and this maximum will have a FWHM of $h\Gamma/(2\pi)$. In reality, the effect of the band structure of the material and the presence of interband transitions can distort this simple free-electron form of the energy loss function. These effects are shown schematically in *Figure 4.5*.

From an experimental low loss spectrum, it is possible via Fourier-Log deconvolution to obtain the SSD and so derive the energy loss function. To obtain this quantity on an absolute scale, it is necessary to normalize the data using Sum Rule techniques – which rely on integrating the energy loss function up to a given energy loss and setting the value of the integral to the total number of valence electrons in the solid. From the

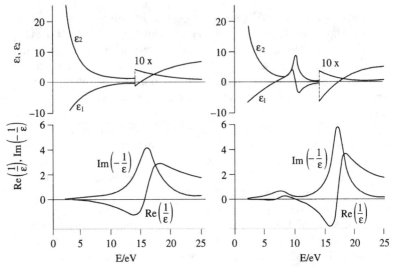

**Figure 4.5.** Schematic diagram of the dielectric properties of (a) a free electron gas and (b) a free electron gas with low energy interband transitions. Redrawn from Daniels *et al.* (1970) *Springer Tracts in Modern Physics,* Vol. 54, with permission of Springer-Verlag New York Inc.

correctly normalized energy loss function, it is possible, using a technique known as Kramers–Kronig analysis, to derive the real and imaginary parts of $\varepsilon$ ($\varepsilon_1$ and $\varepsilon_2$ respectively). A zero crossing of $\varepsilon_1$ (with a positive slope) corresponds to a plasmon oscillation, while peaks in $\varepsilon_2$ correspond to interband transitions which may be analysed in terms of a JDOS (a convolution) between the valence band DOS and conduction band DOS. Sum rule techniques can determine the number of valence electrons involved in each of the separate excitations. In recent years quantitative modelling of low loss spectra using these approaches has been achieved and the results directly related to electronic band structure calculations (Mullejans and French, 1996).

## 4.3 Experimental aspects of EELS low loss measurements

The large signal available in the low loss region makes the acquisition of statistically significant data with a good SNR relatively straightforward. In some cases, the high intensity of low loss data, particularly in the narrow ZLP, can lead to saturation of parallel detection systems and lead to unwanted memory effects. These effects should be removed by an accurate dark count correction discussed in Section 3.3.2. Further experimental benefits of low loss spectra include the ability to easily calibrate the absolute energy scale via use of the ZLP, which is by definition the origin of the energy scale.

As discussed in Section 4.1, low loss data is easily deconvoluted using the Fourier-Log method, however, it is wise to measure the data over a sufficiently wide energy range so as to allow the accurate extrapolation of the high energy loss end of the low loss data to zero. For any form of quantitative analysis of low loss data, deconvolution should be routinely employed.

Generally the overall spectral energy resolution and/or correction for the detector point spread function is not of great concern in the low loss region since most features such as plasmons or peaks in the JDOS are relatively broad. However, one area where both energy resolution and the correction for the PSF is important is in the determination of band gap energies, since a narrow ZLP can be more easily extrapolated and removed from the low loss spectrum to reveal the onset of transitions to the conduction band.

## 4.4 Conclusions

To conclude, the low loss region of the spectrum exhibits a relatively high signal and contains much useful information for the solid-state scientist. It provides a relatively quick and semi-quantitative measure of the sample thickness, which can become fully quantitative if the mean free path is accurately known. The plasmon peak position is a function of valence electron density in the material and therefore provides a means of phase identification; furthermore, its sensitivity allows an investigation of the degree of solid solution formation in an alloyed material and also structural and electronic changes. The low loss region can also provide a measure of the band gap of a material, the energy of valence to conduction band transitions and finally, a detailed analysis can give data which can be directly compared and, in some cases extend, UV and visible data.

## References and further reading

**Daniels, J., Festenberg, C.V., Raether, H. and Zeppenfeld, K.** (1970) Optical constants of solids by electron spectroscopy. *Springer Tracts in Modern Physics*, Vol. 54. Springer-Verlag, New York, pp. 78–135.

**Disko, M.M., Ahn, C.C. and Fultz, B. (eds)** (1992) *Transmission Electron Energy Loss Spectrometry in Materials Science*. TMS, Warrendale, Pennsylvania.

**Egerton, R.F.** (1996) *Electron Energy Loss Spectroscopy in the Electron Microscope*. Plenum Press, New York.

**Monthioux, M., Soutric, F. and Serin, V.** (1997) Recurrent correlation between electron energy loss spectra and mechanical properies for carbon fibres. *Carbon* **35:** 1660.

**Mullejans, H. and French, R.H.** (1996) Interband electronic structure of a near Σ11 grain boundary in α-alumina determined by spatially resolved valence electron energy loss spectroscopy. *J. Appl. Phys.* **29:** 1751.

**Raether, H.** (1980) Excitations of plasmons and interband transitions by electrons. *Springer Tracts in Modern Physics*, Vol. 88. Springer-Verlag, New York.

**Rafferty, B. and Brown, L.M.** (1998) Direct and indirect transitions in the region of the band gap using EELS. *Phys. Rev. B* **58:** 10326.

**Ritchie, R.H.** (1957) Plasmon losses by fast electrons in thin films. *Phys. Rev.* **106:** 874–881.

# 5  Elemental quantification

## 5.1 Quantification of EEL spectra

As highlighted in Section 2.3, PEELS in the environment of the analytical TEM enables the direct detection of the presence of differing elemental species in different microstructural regions within a sample. This is achieved via the identification of characteristic ionization edges in the spectrum, which arise due to the excitation of inner-shell electrons in atoms within the sample. Qualitative determination of sample compositions may be achieved via simple comparison of the energies of the observed inner-shell ionization edges in an EEL spectrum with either a table of atomic energy levels (often included in EELS data processing software) or the *EELS Atlas* (Ahn and Krivanek, 1983), which contains a selection of reference spectra. *Figure 5.1* shows EELS spectra for MnO and $Fe_2O_3$ in which the oxygen $K$-edges are at 532 eV while the manganese and iron $L_{2,3}$-edges are at 640 and 710 eV, respectively.

After some initial data processing, such as dark current subtraction and gain correction which are a function of the response of the detection system outlined in Section 3.8, in order to quantify the elemental analysis it is necessary to measure the intensities under the various edges. This is achieved by fitting a background (in this case a power-law $A.E^{-r}$), indicated by the dotted lines, to the spectrum immediately before the edge. This is then subtracted and the intensity is measured in an energy window, $\Delta$, which begins at the edge threshold and usually extends some 50–100 eV above the edge. The next step is to compute the inelastic *partial cross-section*, $\sigma$, for the particular inner-shell scattering event under the appropriate experimental conditions, that is, $\sigma(\underline{\alpha}, \underline{\beta}, \underline{\Delta}, \underline{E_o})$; note the bracketed terms underlined simply represent the variables of the particular function, in this case the partial cross-section. This partial cross-section is used to scale or normalize the measured edge intensity so that either different edge intensities can be compared, or the intensity can be directly interpreted in terms of an atomic concentration within the specimen volume irradiated by the electron probe.

EELS quantification procedures, although well established, are relatively user intensive and are somewhat dangerous if excessively

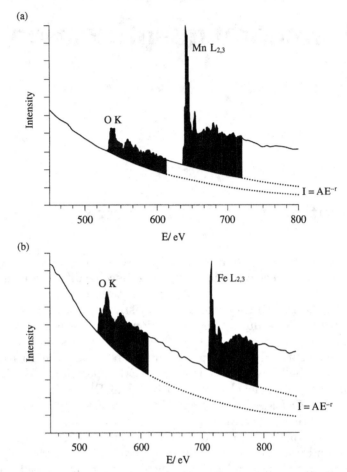

**Figure 5.1**. EELS spectra of (a) MnO and (b) $\alpha$-$Fe_2O_3$. In both cases fitted power law ($I = A.E^{-r}$) backgrounds are indicated by dotted lines. The shaded areas extend for an energy window, $\Delta = 80$ eV, above each edge threshold and are used to quantify the elemental analysis as described in the text.

over-automated. For accurate analysis, typically EEL spectral quantification involves a number of stages, which are discussed in more detail below.

## 5.2 Background removal

Edge quantification requires the determination of the intensity due to solely the ionization events. This is achieved by the subtraction of the continuously decreasing background component from the edge signal of each element of interest. All core loss edges lie on a background composed of the tails of the plasmon excitation(s) plus those from ionization edges with lower threshold energies. This relatively large background contribution is one of the major disadvantages of EELS elemental quantification, making weak edges difficult to see and analyse. If the SNR is adequate, the use of a small collection angle will increase the SBR as discussed in Section 3.5.2. Fourier-Log deconvolution to remove multiple inelastic scattering will make also make edges more readily visible above the background as outlined in Section 4.1.

In high-energy regions away from the low loss regime where the plasmon contribution is dominant, and sufficiently beyond any other edge threshold, there is an approximately linear relationship between the logarithm of the intensity and the logarithm of the energy loss. This *log–log relationship* means that the background can be fitted using a power-law function of the form $I=A.E^{-r}$. There are two main ways to evaluate the fitting parameters $A$ and $r$:

(i) Most commonly a linear least squares fit is made to a suitable region of the $\log(I)$–$\log(E)$ plot on a computer, this region is usually chosen to be directly prior to the ionization edge onset. The resulting background fit is then extrapolated under the edge of interest. The choice of fitting and extrapolation regions is dependent on the system, but these should normally be similar and also be as large as possible bearing in mind the constraints induced by the presence of neighbouring edges. *Figure 5.2* shows typical fitting, extrapolation and integration regions for the background under a nitrogen $K$-edge.

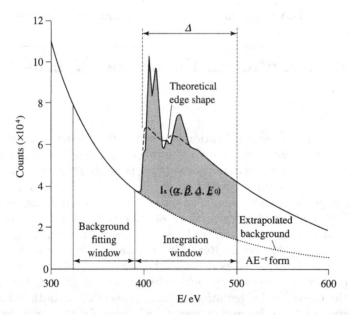

**Figure 5.2.** The nitrogen $K$-edge from thin sample of boron nitride with a power law background fitted in a window prior to the edge and extrapolated over a window of width, $\Delta$, under the edge. Also shown is a theoretical N K-edge cross-section calculated using the hydrogenic model after convolution with the low loss region of the spectrum to simulate the effects of multiple inelastic scattering.

(ii) Another technique is the so-called *two-area method*, which involves measuring the areas (intensities) in two windows of equal widths prior to the edge (termed pre-edge windows). Assuming the power law dependence of the background, an algorithm can then be used to extract the background intensity over a window of equal width under the ionization edge of interest (the post-edge window). This technique, described more fully in Egerton (1996), is often used in the processing of ESI images to produce quantitatve maps of elemental concentrations (see Chapter 7).

Often, due to the presence of neighbouring edges or a strong plasmon contribution, the spectral background deviates significantly from the ideal log–log behaviour. Several techniques have been used to overcome this problem. In many cases the $\log(I)$–$\log(E)$ fit is constrained, either to pass through a specified point at much higher energy loss, or to follow the form of the ionisation edge above the edge threshold. In the low-energy loss region, sometimes a log-linear (i.e. $\log(I)$ vs. $E$) fitting function performs reasonably well, while another possibility is to accurately model the plasmon shapes for a specimen of the appropriate thickness and use these in addition to an $A.E^{-r}$ background function. Further problems may arise from the accurate separation of overlapping, closely spaced edges. In the latter case, the spectrum is often fitted to a series of reference spectra using a multiple least-squares fitting procedure – fitting the differentiated or double-differentiated spectrum can often remove the problem of the slowly varying background (Leapman, 1992).

---

# 5.3 Determination of the ionization cross-section

---

The next step in the quantification process is the determination of the partial ionization cross-sections for the relevant ionization edges, either in absolute terms or relative to those other elements present. Knowledge of the relevant partial cross-sections allows the conversion of the extracted edge intensities into elemental concentrations. The absolute values of the ionization cross-sections may be calculated using either a simple hydrogenic model (Egerton, 1996), or a more complicated Hartree–Slater model for the atom (Rez, 1982); alternatively, relative cross-sections may be determined experimentally via measurement of standard samples of known composition such as binary oxides (Hofer, 1987). The basis for the theoretical calculation of the cross-sections is outlined below.

To describe the angular and energy distributions for scattering from an atom into a solid angle $d\Omega$ with an energy loss between $E$ and $E+dE$, it is normal to use the double differential cross-section, $d^2\sigma/(dE\ d\Omega)$, introduced in Section 2.4. This quantity is a function of incident beam energy, $E_o$. Since the double differential cross-section is symmetric about the incident beam direction, a spectrometer collecting a cone of scattering with semi-angle, $\beta$, about the forward direction yields a partial differential cross-section which is also a function of $\beta$:

$$d\sigma(\beta, \underline{E_o})/dE = \int_0^\beta d^2\sigma(\underline{E_o})/(dE\ d\Omega).2\pi\ \sin\theta\ d\theta \qquad (5.1)$$

Again the terms underlined represent the variables of the particular function. Diffraction effects can modify this integral as they may destroy the cylindrical symmetry contained in the integral over the azimuthal angle which is equal to $2\pi$ for the symmetric case.

After extrapolation and removal of the preceding background, the edge signal is integrated over a suitable energy window, $\Delta$, above the edge threshold, $E_t$. This energy window is usually chosen to be sufficiently large (e.g. $\Delta > 50$ eV) so as to minimize the effects due to the solid-state character of the specimen which results in the characteristic ELNES observed at the ionization edge (see Chapter 6). The appropriate partial cross-section is then

$$\sigma(\underline{\beta}, \underline{\Delta}, \underline{E_o}) = \int_{E_t}^{E_t+\Delta} d\sigma(\underline{\beta}, \underline{E_o})/dE\ dE \qquad (5.2)$$

If the incident beam is convergent within a cone of semi-angle, $\alpha$, the differential cross-section must be convoluted with the incident angular distribution before integration over the collection angle (Equation 5.1). The partial cross-section then also becomes a function of $\alpha$. This is only of major importance if $\alpha > \beta$ (i.e. for highly convergent probes), as possibly in STEM or nanoprobe CTEM (Egerton, 1996).

For a given set of experimental parameters, as a rough guide, the cross-section depends mainly on the number of inner shell electrons divided by the energy loss (Equation 1.7). Many cross-sections may be determined using simple routines, such as Egerton's *SIGMAK2* and *SIGMAL2* hydrogenic programs (Egerton, 1996), installed on personal computers or embedded within commercial EELS data processing packages. An alternative method is to employ *k-factors*, similar to those used in X-ray analysis of thin films, derived from measurements on standards of known chemical composition under the same experimental conditions (Hofer, 1987; Hofer and Kothleitner, 1993, 1996). These *k*-factors essentially represent ratios of partial ionization cross-sections for a particular set of experimental parameters $\underline{\alpha}$, $\underline{\beta}$, $\underline{\Delta}$ and $\underline{E_o}$.

To calculate the partial ionization cross-sections explicitly, it is necessary to form an expression for the double differential cross-section and then integrate according to Equations (5.1) and (5.2). From Section 2.4, Bethe theory allows us to write the double differential cross-section as the product of two terms: one dependent of the incident electron and the other dependent only on the excited atom.

$$d^2\sigma/(dE\ d\Omega) = 4\gamma^2 R/(Ek_0^2)\ 1/(\theta^2 + \theta_E^2)\ df/dE \qquad (5.3)$$

To recap, here $\gamma$ is the relativistic correction factor for a particle of velocity, $v$, given by Equation (1.1), $R$ is the Rydberg=13.61 eV, $k_0$ is the incident wave vector, $E$ is the energy loss, $\theta$ is the scattering angle and $\theta_E$ is the characteristic angle $= E/(\gamma m_e v^2) \approx E/(2E_0)$. The atom-dependent part is known as the GOS per unit energy loss, $df/dE$, which depends both on energy loss, $E$, and the change in wavevector, $q$ given by Equation (2.1). In Bethe theory the GOS term may be written as

$$df(q,\ E)/dE = E/(Ra_o^2 q^2).\sum | <E_{\text{final}}\ l'| \exp\ (i\mathbf{q.r})|nl> |^2 \qquad (5.4)$$

where $a_o$ is the Bohr radius = 0.053 nm; $|nl>$ is the initial state of the atomic electron with principle quantum number, $n$, and angular momentum quantum number, $l$; $<E_{\text{final}}l'|$ is the final unbound state of the excited electron with energy $E_{\text{final}}$ and angular momentum quantum number $l'$, while $r$ is the position of the atomic electron. The second term in Equation (5.4) is known as the square of the transition matrix element between the initial state of the atomic electron and a particular unoccupied final state. Note, Equation (5.4) is a reformulation of Equation (2.6) where the individual integrals are now an explicit sum over all possible final states. This arises because we have to sum over all the individual cross-sections for each of the transitions. As discussed in Section 6.1, not all final states, and hence transitions, will be allowed if the dipole selection rule is obeyed during the recording of the experimental data (i.e. measurements made with small $q$ and hence small collection semi-angle); the allowed final states will be those where $l' = l \pm 1$. The form of the GOS for the carbon $K$-shell is shown in *Figure 2.3* (Section 2.4).

Currently there are two methods for calculating the GOS, which can then be used to calculate the partial scattering cross-sections used for quantification. Both these methods do not incorporate transitions to bound states but simply sum over all unbound final states in Equation (5.4); this provides a free atom cross-section and consequently the detailed ELNES shapes are not predicted. The first method uses simple hydrogenic-like wave functions for the calculation of $K$ and $L_{2,3}$ cross-sections (the SIGMAK2 and SIGMAL2 programs of Egerton, 1996). The second uses the more realistic Hartree–Slater–Fock wavefunctions and have been used for $K$-, $L_1$-, $L_{2,3}$- and even $M_{2,3}$- and $M_{4,5}$-edges (Egerton, 1996; Rez, 1982). A parameterization scheme for partial cross-sections has also been proposed (Egerton, 1993). *Figure 5.2* shows the theoretical edge shape for the N $K$-edge calculated using hydrogenic wavefunctions. Note the absence of the experimentally observed ELNES in the theoretical shape derived using the free atom model.

The use of free atom cross-sections for quantification can be justified if the differential cross-section in Equation (5.2) is integrated over a sufficiently large enough energy window, $\Delta$. In other words, the solid state environment of the atom, encapsulated in the ELNES, essentially redistributes the energy of these free atom states but does not create or remove states. Problems can occur for particular edges which exhibit intense transitions to bound states, such as the $L_{2,3}$- and $M_{4,5}$-edges of the transitions metals and rare earths which show sharp 'white lines' at the edge onset, reflecting transitions to tightly bound $d$ or $f$ states. Here the calculated cross-sections are often empirically corrected for the presence of these white lines using photoabsorption data (Egerton, 1996).

# 5.4 Final quantification step

Once the edge intensity has been extracted and the relevant partial cross-section been either calculated or measured for a given set of experimental parameters ($\underline{\alpha}$, $\beta$, $\underline{\Delta}$ and $\underline{E}_o$), it is then possible to derive the concentration of a particular element X within the specimen area irradiated by the electron beam (the so-called areal density). To do this we assume that the inner-shell signal from a specimen of thickness $t$, containing $n_X$ atoms per unit volume is given by

$$I_X(\underline{\alpha},\ \beta,\ \underline{\Delta},\ \underline{E}_o) = I_{ZL} . \sigma(\underline{\alpha},\ \beta,\ \underline{\Delta},\ \underline{E}_o) . n_X . t \tag{5.5}$$

where $I_{ZL}$ is the ZL signal. Strictly, Equation (5.5) applies to an edge which has been deconvoluted for the effects of multiple inelastic scattering (Section 4.1). To take into account the effects of thickness, which will transfer the intensity in both the edge signal and the low loss signal to higher energies via multiple inelastic events, it is more accurate to integrate the total intensity of the ZL and low loss signals, $I_{LL}$, over the same energy window $\Delta$, giving the areal density of element X, $n_X.t$, as

$$n_X.t = I_X(\underline{\alpha},\ \beta,\ \underline{\Delta},\ \underline{E}_o)/[I_{LL}(\underline{\alpha},\ \beta,\ \underline{\Delta},\ \underline{E}_o) . \sigma(\underline{\alpha},\ \beta,\ \underline{\Delta},\ \underline{E}_o)]. \tag{5.6}$$

Equation (5.6) therefore permits the determination of the *absolute atomic concentrations* of elemental species, if the sample thickness is accurately known; the probe size and possibly probe broadening are also required if *absolute numbers of atoms* are to be obtained. However, as shown in *Figure 5.1*, experimentally the more usual procedure is to determine *relative atomic concentration ratios* of various combinations of elements such as $n_O/n_{Fe}$ and $n_O/n_{Mn}$. Here, provided the same integration window is employed, the low loss intensity and the specimen thickness cancel and, for the case of MnO, the atomic concentration ratio is given by:

$$n_O/n_{Mn} = \{\sigma_{Mn}(\underline{\alpha},\ \beta,\ \underline{\Delta},\ \underline{E}_o)/\sigma_O(\underline{\alpha},\ \beta,\ \underline{\Delta},\ \underline{E}_o)\} . \{I_O(\underline{\alpha},\ \beta,\ \underline{\Delta},\ \underline{E}_o)/I_{Mn}(\underline{\alpha},\ \beta,\ \underline{\Delta},\ \underline{E}_o)\}$$
$$(5.7)$$

where, for a given set of experimental parameters ($\underline{\alpha}$, $\beta$, $\underline{\Delta}$ and $\underline{E}_o$), $I_O$ and $I_{Mn}$ are the measured intensities under the oxygen and manganese core-loss edges and $\sigma_O$ and $\sigma_{Mn}$ are the appropriate partial cross-sections for the O $K$- and Mn $L_{2,3}$-edges.

## 5.5 Summary of the quantification procedure

(i) Correct spectrum for artefacts, that is, dark current, diode response (PEELS), gain change (serial EELS).
(ii) Deconvolute the spectra using a corresponding low loss spectrum from the same specimen area (Section 4.1, this is optional, but recommended for all but thinnest samples).
(iii) Select the fitting region, extrapolation range and width of the integration window to be used on all edges of interest.
(iv) Integrate the low loss signal over the same integration window as chosen in (iii) if absolute concentrations are required.
(v) Fit a background prior to the edge and subtract the extrapolated background intensity from the edge signal.
(vi) Integrate the edge signal(s) over the chosen integration window.
(vii) Calculate the partial cross-section(s) for the particular set of experimental parameters employed.
(viii) Use Equation (5.6) to calculate the number of atoms per unit sample area, or Equation (5.7) to determine the ratio of the numbers of atoms in the probed volume.

## 5.6 Summary of experimental quantification parameters

For accurate quantification it is important to know the following experimental parameters:

(i) Incident beam energy, $E_o$.
(ii) The collection semi-angle, $\beta$.
(iii) The convergence semi-angle, $\alpha$
(iv) The energy window for edge integration, $\Delta$.

The values of the collection and convergence semi-angles may be determined using the procedures outlined in Section 3.5.1. All these parameters need to be specified in the routines used to calculate the partial cross-sections, or if standards are to be used, the $k$-factors need to be derived from measurements on standards using identical experimental conditions.

For quantitative analysis, as a general rule it is advisable to avoid specimen orientations which give rise to strongly diffracting conditions, as the presence of Bragg diffraction spots and channelling effects can significantly affect the accuracy of the equations derived for quantitative analysis in Section 5.4 and lead to anomalous results. Furthermore, for accurate results the specimen thickness should be less than $\Lambda$, the mean free path for inelastic scattering, and preferably around half this value; spectra from thicker regions should be deconvoluted using the corresponding low loss spectrum. Note that extremely thin specimen regions ($t/\Lambda < 0.2$) may possess modified surface layers a few nanometres in thickness (e.g. surface oxides) which could represent a large proportion of the overall probed volume and will lead to erroneous conclusions about atomic concentration ratios of particular elements; in practice, sets of measurements from regions of differing specimen thickness can identify and remove such effects.

## 5.7 Accuracy and detection sensitivity of EELS quantification and comparison with EDX in the TEM

For accurate quantification, it is important that spectra deconvoluted for the effects of thickness are employed. Ignoring any specimen-induced inaccuracies (due to excessive sample thickness, strongly diffracting orientation or surface layers etc.), the overall accuracy mainly depends on the accuracy of the calculated or derived cross-section as well as the background fit and extrapolation. If each of these stages is optimized, atomic concentration values typically within 5% of the true composition have been shown to be attainable (Egerton, 1996; Leapman, 1992). Generally relative atomic concentrations of selected elements are determined, although absolute concentrations can be obtained provided the size of the sampled volume is known which will require an accurate knowledge of the sample thickness (Section 4.1).

Typically EELS detection sensitivities for particular elements lie in the range 0.1 to 1 atom% but vary greatly with specific experimental factors as well as the mix of elements present in a given sample (Leapman, 1992). PEELS is significantly better than EDX analysis for light elements ($Z < 11$) and for many transition metal and rare earth elements, and *Figure 5.3* shows a comparison of the detection sensitivities of EELS and EDX as function of atomic number. However, in the TEM environment, the two forms of quantitative microanalysis should definitely be employed in combination since EDX is undoubtedly superior for heavier elements as the probability of de-excitation by X-ray emission over Auger electron emission increases with atomic number $Z$.

EDX provides a much wider spectral energy range for the collection of statistically significant data (typically 20 keV compared with 2 keV in the case of EELS) and therefore permits access to a larger array of core level ionization signals. However, at present (in the absence of widespread use of new microcalorimeter EDX detectors), the spectral resolution in EDX is significantly worse than in EELS and can lead to severe problems for compounds containing both light and heavy elements when overlapping X-ray lines and self-absorption of soft X-rays render quantitative analysis extremely difficult.

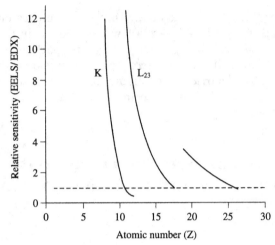

**Figure 5.3.** Comparison of the relative sensitivities of EELS and EDX (i.e the ratio of the SNRs) as a function of atomic number.

For high spatial resolution analysis, EDX is impaired by the effects of beam broadening even in thin foils (which limits the ultimate analytical resolution to at best a tens of nanometres); this is absent in most EELS experiments since it is usual to limit the analytical volume sampled via use of a well-defined collection aperture. The quantity of most relevance in high spatial resolution microanalysis is the minimum detectable mass of a particular element, which depends critically on the electron probe size and probe current. In STEMs and modern field emission TEMs, the minimum detectable mass using EELS can, for many elements in a thin, low atomic number matrix, approach 1–10 atoms (Egerton, 1996).

# References and further reading

**Ahn, C.C. and Krivanek, O.L.** (1983) *EELS Atlas*. ASU Centre for Solid State Science, Tempe, Arizona and Gatan Inc., Warrendale, Pennsylvania.

**Egerton, R.F.** (1993) Oscillator strength parameterization of inner shell cross-sections. *Ultramicroscopy* **50:** 13–28.

**Egerton, R.F.** (1996) *Electron Energy Loss Spectroscopy in the Electron Microscope*. Plenum Press, New York.

**Hofer, F.** (1987) EELS quantification of *M* edges using oxidic standards. *Ultramicroscopy* **21:** 63–68.

**Hofer, F. and Kothleitner, F.** (1993) Quantitative microanalysis using EELS: I. Li and Be in oxides. *Microsc. Microanal. Microstruct.* **4:** 539–560.

**Hofer, F. and Kothleitner, G.** (1996) Quantitative microanalysis using EELS: II. Compounds with heavier elements. *Microsc. Microanal. Microstruct.* **7:** 265–277.

**Leapman, R.D.** (1992) EELS quantitative analysis. In: *Transmission Electron Energy Loss Spectrometry in Materials Science* (eds M.M. Disko, C.C. Ahn and B. Fultz). TMS, Warrendale, Pennsylvania.

**Rez, P.** (1982) Cross-sections for energy loss spectrometry. *Ultramicroscopy* **9:** 283–288.

# 6 Fine structure on inner-shell ionization edges (ELNES/EXELFS)

## 6.1 Origin of edge fine structure

As we have seen in Sections 2.4 and 5.3, EELS inner-shell ionization edges may be modelled via the calculation of the double differential scattering cross-section. This, in turn, can be achieved by calculation of the GOS per unit energy loss, $df/dE$. A relatively simple method is to assume a free-atom model, summarized in Equations (2.8), (5.3) and (5.4). This model predicts the basic shape of the inner-shell ionization edges employed in the elemental quantification routines described in Chapter 5; however, it is clear from *Figure 5.2* that the detailed fine structure exhibited by the ionization edge is not reproduced. As we will see, this fine structure is a direct result of the specific local environment of the atom undergoing ionization and can give direct insight into the associated electronic structure and bonding in the solid.

In the free atom model, we shall see that the GOS depends on the real space overlap between the initial core level wavefunction and all the various atomic-like final states coupled by the *dipole selection rule*. This selection rule governing the observed electronic transitions originates from the expansion of the exponential term in Equation (2.6).

$$\exp(i\mathbf{q}.\mathbf{r}) = 1 + i\mathbf{q}.\mathbf{r} + \dots . \tag{6.1}$$

For small $q$, we can neglect any higher power terms, such as those involving $(\mathbf{q}.\mathbf{r})^2$, in the mathematical expansion of this exponential function as a series. Since we square these terms in the expression for the GOS in Equation (2.6), the latter forbidden terms are known as quadrupole terms and may be significant if large collection angles (typically $> 30$–$40$ mrads) are employed during experimental measurement. More accurately the dipole approximation is valid if $\mathbf{q}.\mathbf{r} < < 1$ for all $r$, which is equivalent to $q < < 1/r_i$ where $r_i$ is the radius of the initial state wavefunction. In the hydrogenic model, $r_i \approx a_0/Z^*$, where $a_0$ is the Bohr radius and $Z^*$ is the effective (screened) nuclear charge as defined by Slater (Egerton, 1996). Using the dipole approximation, we can simplify part of the expression for the GOS in Equations (2.6) and (5.4)

$$|<f \mid \exp(i\mathbf{q}.\mathbf{r})\mid i>|^2 = |<f \mid \mathbf{q}.\mathbf{r}\mid i>|^2 \tag{6.2}$$

This simplification arises because the two wavefunctions, $\Psi_i$ and $\Psi_f$ are mutually orthogonal (essentially this means that they are independent functions which is a fundamental condition for the allowed solutions of the Schrödinger equation); the integral involving the first term in the expansion (i.e. 1) will therefore be zero leading to Equation (6.2). The dipole approximation will also affect the nature of the initial and final states, in particular their symmetry in terms of angular momentum (e.g. $s, p, d, f, \ldots$). If the initial and final state wavefunctions are of the same symmetry (i.e. either both even or both odd functions), then the integral will be zero. For the integral to be non-zero, the wavefunction symmetry has to change during the electronic transition, such as a transition from an $s$ state (an even function) to a $p$ state (an odd function). This condition leads to the dipole selection rule for allowed transitions being formulated as $\Delta l = \pm 1$, where $\Delta l$ *is* the change in angular momentum during the transition from the initial to the final states. This is exactly the same rule as is observed for electronic transitions produced via photon-based spectroscopies such as X-ray emission following inner-shell ionization.

Thus from Chapter 5, the energy differential cross-section, under the dipole selection rule, is given by

$$d\sigma/dE = \int d^2\sigma/(dE \, d\Omega) \, d\Omega$$
$$= \int 4\gamma^2/(a_0^2 q^4) \, (k_f/k_0) \, \Sigma_f \mid <f\mid \mathbf{q}.\mathbf{r}\mid i>\mid^2 \, d\Omega \tag{6.3}$$

Note here we have summed over all possible final states, which is identical to summing over all possible differential cross-sections for each of the transitions. We can expand the integral in Equation (6.3) as

$$|<f\mid \mathbf{q}.\mathbf{r}\mid i>|^2 = |<f\mid q_x x + q_y y + q_z z \mid i>|^2 = q^2 \mid <f\mid \mathbf{r}\mid i>\mid^2 \tag{6.4}$$

We can ignore any cross terms (involving $q_x q_y$, etc.) in the square of the integral since, for the case of a circular collection aperture, they will disappear in the integration of Equation (6.3) over the solid angle.

Thus the spatial overlap between the initial and final states of the excited electron governs the magnitude of the energy differential cross-section for the transition. This overlap will be similar for each particular set of initial and final states that give a non-zero integral and are thus allowed by the dipole selection rule. This leads to the concept of a set of basic shapes for each class of inner-shell ionization edge which may be calculated using the free atom model outlined in Section 5.3. These free atom edge shapes are schematically displayed in *Figure 6.1*. For instance, where the initial atomic core state is an $s$ state, the dipole allowed final states will be of $p$ symmetry; these transitions are observed at $K$-, $L_1$- and

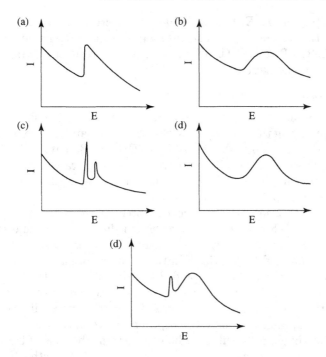

**Figure 6.1.** Schematic diagram of the basic shapes of EELS ionization edges as predicted by the free atom model with the addition of bound states in (c) and (e): (a) sawtooth shape, e.g. $K$-edges; (b) delayed maximum, e.g. $L_{2,3}$-edges of third and fourth period elements and $M_{4,5}$-edges of fifth period elements; (c) white line, e.g. $L_{2,3}$-edges of $3d$ and $4d$ transition metals, $M_{4,5}$-edges of rare earths; (d) plasmon-like, e.g. $M_{2,3}$-edges of fourth period elements; (e) mixed shape, e.g. bound state plus delayed maximum.

$M_1$-edges and give rise to a characteristic 'sawtooth' shape. Meanwhile, dipole transitions from an initial core level of $p$ symmetry, such as $L_{2,3}$-edges and $M_{2,3}$-edges exhibit a delayed maximum, except in cases such as the transition and rare-earth elements where the high density of unoccupied bound $d$- or $f$-states are important and give rise to a intense 'white-line' feature near the edge threshold superimposed on this delayed maximum shape.

In practice, it is observed that experimental inner-shell ionization edge spectra from solid materials exhibit a considerable amount of detailed fine structure superimposed on the basic atomic-like shape described above. This structure is termed *ELNES* and *EXELFS*; ELNES is the most intense and is apparent within the first 30–40 eV above the edge onset (Egerton, 1996). As discussed in Section 1.2.1, this fine structure arises because in solids the atoms are not free but their outer electrons overlap with those on neighbouring atoms and are involved in inter-atomic bonding. Thus the detailed shape and form of the ELNES is a solid state effect which cannot be predicted by the free-atom model.

What is required in Equation (6.3) is not just the sum over all the individual cross-sections from the initial state to all possible final states of a particular energy $E$, but each individual cross-section needs to be weighted by the number of final states available at that particular energy (Brydson, 1991; Rez, 1991). We can then express the total energy differential cross-section as

$$d\sigma/dE \propto \Sigma_f \mid M_f(E)\mid^2 . N_f(E) \tag{6.5}$$

where $M(E)$ is an atomic transition matrix (a set of transition matrix elements) governing the overlap between the initial and all the individual final states (the basic edge shape), while $N(E)$ is the density of final states (DOS term), the number of final states in an energy range of $E$ to $E+dE$ which was first introduced in the discussion of electron energy band theory in Section 1.2.1.

The atomic transition matrix governs which subset of the total density of states is probed. For $M_f$ to be non-zero and the transition to occur, the final state needs to be unoccupied; it also needs to have appreciable overlap with the highly localized initial atomic core state. Thus $N(E)$ effectively represents an unoccupied, local density of states (i.e. empty states in the conduction band of the material which have appreciable weight on the atomic site being ionized). Furthermore, if the dipole approximation applies (i.e. for small $q$), then, as we have seen, the atomic transition matrix element will only be non-zero for the case of transitions in which the change in angular momentum, $\Delta l$ is $\pm 1$. Thus the $N(E)$ term in Equation (6.5) is in fact a local, symmetry-projected unoccupied DOS, not a total DOS term. One commonly used approximation is to assume that the dipole allowed transition matrix elements vary smoothly as a function of energy, implying that the ELNES directly reflects this local symmetry-projected DOS term. A schematic, simplified diagram of how the DOS term is reflected in the ELNES is shown in *Figure 6.2*.

The local site-projection of the unoccupied DOS is reflected in the different edge shapes observed for different elements in a particular

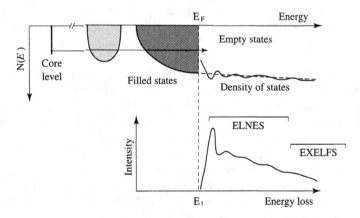

**Figure 6.2.** Schematic diagram showing how the ELNES intensity above the edge threshold, $E_t$, reflects the unoccupied DOS probed by an inner-shell electron excited from a deep core level.

compound, demonstrated in *Figure 6.3* for the Si *K*- and O *K*-edges in crystalline $SiO_2$ quartz (Garvie *et al.*, 2000). Meanwhile, the symmetry-projection results in different inner-shell edges of the same element probing differing symmetries of the unoccupied DOS; in *Figure 6.3* the Si *K*- and Si $L_1$-edges in quartz both probe the unoccupied *p*-like DOS local to the silicon atom, whereas the Si $L_{2,3}$-edge probes the local unoccupied *s*- and *d*-like DOS resulting in very different ELNES.

Finally, until now we have been considering an initial core level state, which is essentially an energy delta function (i.e. it is a single, strongly bound atomic level of negligible energy width). If the initial state is more weakly bound and hence more band-like, we also have to consider the energy density of initial states and $N(E)$ then becomes a convolution between the initial and final densities of states, known as a joint DOS or JDOS. This is really only applicable in the low loss regime which includes transitions originating from weakly bound inner shells, these may be thought of as a form of interband transition as discussed in Section 4.2.

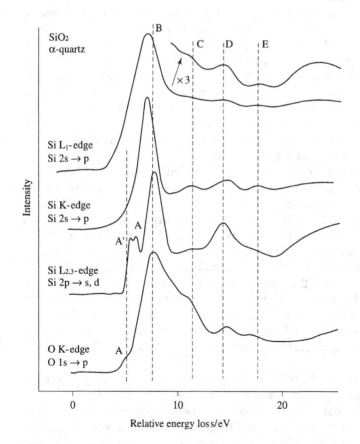

**Figure 6.3.** Background subtracted inner-shell ionization edges from α-$SiO_2$ quartz: Si $L_1$-, Si *K*-, Si $L_{2,3}$-, and O *K*-edges demonstrating the site and symmetry selectivity displayed by ELNES.

Returning to Equation (6.5), we may summarize the situation by stating that in a solid, the energy differential cross-section for inner-shell ionization, and therefore the detailed ionization edge shape, will be proportional to a site- and symmetry-projection of the unoccupied DOS. Since the exact form of both the occupied and unoccupied DOS will be appreciably modified by the presence of bonding between atoms in the solid, this will therefore be reflected in the detailed ionization edge structure known ELNES. We now discuss how ELNES measurements can in many cases, allow the semi-quantitative extraction of such bonding information.

## 6.2 Determination of coordinations

As we have seen, the exact form of the differential cross-section and therefore the ELNES on a particular core-loss edge is highly dependent on the local atomic environment around the atom undergoing excitation. It is this environment which determines the density of unoccupied electron states projected onto the particular atom in question. This has been demonstrated in *Figure 6.3* for the silicon and oxygen $K$-edges in $SiO_2$ quartz. Under suitable experimental conditions, both edges arise from dipole transitions from the respective atomic $1s$ core levels to unoccupied $p$-like states in the conduction band; however, the local nature of the DOS term causes the two atomic sites to exhibit distinctly different ELNES. In principle this allows us to determine qualitatively the relative atomic environments in a solid.

In many cases, it is found that, for a particular elemental ionization edge, the observed ELNES exhibits a structure that is specific to the arrangement, i.e. the number of atoms and their geometry, as well as the type of atoms within solely the first coordination shell. This occurs whenever the local DOS of the solid is dominated by atomic interactions within a molecular unit, and is particularly true in many non-metallic systems such as semiconducting or insulating metal oxides where, as discussed in Section 1.2.1, we can often envisage the energy band structure as arising from the broadened molecular orbital levels of a giant molecule. If this is the case, we then have a means of qualitatively determining nearest neighbour coordinations using characteristic ELNES shapes or *coordination fingerprints* (Brydson *et al.*, 1992a; Hofer and Golob, 1988).

This concept of an ELNES fingerprint of the local coordination is illustrated in *Figure 6.4* for the case of the B $K$-edges of boron trigonally and tetrahedrally coordinated to oxygen in two borate minerals (Sauer *et al.*, 1993). The planar trigonal coordination of boron to oxygen results in both $\pi$ and $\sigma$ bonding leading to the presence of transitions to a low-energy, unoccupied $\pi$ antibonding ($\pi^*$) feature in the ELNES in addition

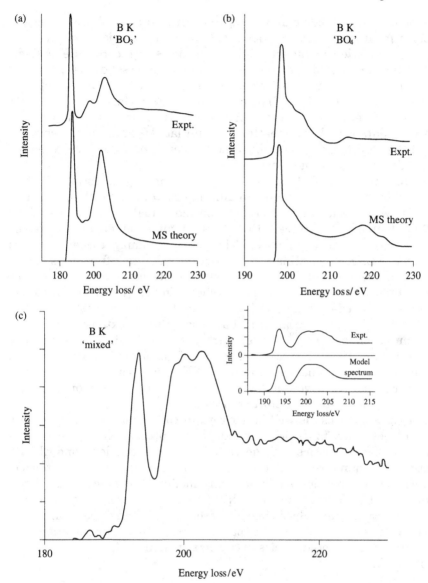

**Figure 6.4.** Comparison of the B $K$-ELNES from: (a) the mineral vonsenite containing trigonal planar $BO_3$ groups; (b) the mineral rhodizite containing tetrahedral $BO_4$ groups and (c) a mixed coordination boron-doped Fe,Cr oxide. In (a) and (b) the experiment (*Expt.*) is compared to the results of single shell MS calculations (*MS Theory*) indicating the existence of trigonal and tetrahedral B $K$-ELNES coordination fingerprints. Inset in (c), the experimental data has been modelled using a linear combination of the B $K$-coordination fingerprints in the ratio 3:1 (trigonal:tetrahedral).

to higher-energy transitions to σ antibonding (σ*) states. This distinct π* feature is absent for the case of boron tetrahedrally coordinated to oxygen where there is solely σ bonding. These edge shapes are common to all boron–oxygen compounds and are even found in boron–nitrogen analogues such as cubic and hexagonal boron nitride. A wide range of cations in different coordinations (e.g. aluminium, silicon, magnesium, various transition metals etc.) and anion units (e.g. carbonate, carbide, sulphate, sulphide, nitrate, nitride etc.) in inorganic solids also show this behaviour, which can be of great use in phase identification and local structure determination (Brydson *et al.*, 1992a; Garvie *et al.*, 1994). Besides reference data present in the literature, an electronic ELNES/ XANES database has recently been initiated (*http://www.cemes.fr/ ~eelsdb/*) which can, in part, be used as a catalogue of both coordination and valence fingerprints (Section 6.3).

Since the inner-shell excitation process is highly localized on a particular atomic site, where two differing coordinations of one element co-exist within the same structure, the individual contributions of the different sites to the observed ELNES is a simple linear sum weighted by the respective site occupancies. Therefore, assuming the existence of distinct ELNES coordination fingerprints, it is often possible to determine semi-quantitatively the relative site occupancies via simple algorithms or fitting routines. This is shown in *Figure 6.4c* for the case of boron in a mixed-coordination boron oxide (Garvie *et al.*, 1995; Sauer *et al.*, 1993). Similar techniques have been used extensively in the determination of $sp^2/sp^3$ bonded carbon ratios in diamond-like films (Berger *et al.*, 1988). Increasing amounts of distortion in structures lead to the broadening and splitting of ELNES features in a coordination fingerprint and, in favourable cases, it may be possible to qualitatively determine the degree of distortion present.

For certain edges, as well as for certain compounds, the concept of a local coordination fingerprint breaks down. In these cases, the unoccupied DOS cannot be so simply described on such a local level and ELNES features are found to depend critically on the arrangement of the atoms in outerlying coordination shells allowing medium range structure determination. This opens up possibilities for the differentiation between different structural polymorphs (Brydson *et al.*, 1992b), as well as the accurate determination of lattice parameters, vacancy concentrations and substitutional site occupancies in complex structures (Kurata *et al.*, 1993; Scott *et al.*, 2001).

## *6.2.1 ELNES modelling*

In order to extract detailed information from ELNES measurements and to determine bonding effects in a more quantitative fashion, some form of accurate spectral modelling is essential. Such modelling entails the computational calculation of the various final state wavefunctions in the solid and their associated energies and, ultimately, the transition matrix and, more importantly, the DOS terms in Equation (6.5).

The calculation of the site- and symmetry-projected unoccupied DOS in the solid may be achieved at varying levels of sophistication (Brydson, 1991; Rez, 1992). Three widely used methods are:

(i) *Molecular orbital (MO) theory* (essentially a chemist's viewpoint) which considers the available electron states formed by a small (often single-shelled) cluster or molecular unit designed to mimic locally the structure of the solid.

(ii) *Band theory* (essentially a physicist's viewpoint) which considers the available electron states formed by an infinitely repeating crystalline lattice (known as Bloch states).

(iii) *Multiple scattering (MS) theory*, which is somewhat intermediate between the other two methods and uses a large cluster to represent the solid. MS theory pictures the ELNES as arising from multiple elastic scattering (i.e. diffraction) of the excited electron by the surrounding atoms in a solid, the outgoing electron wave interacts with the backscattered wave and produces an interference pattern (the ELNES) which is a function of energy of the excited electron.

A further modelling method, known as atomic multiplet theory, is discussed in Section 6.3. However, this method is based on an extremely local description and its applicability is largely confined to certain ionization edges that exhibit intense white-line features, as shown in *Figure 6.1*.

Examples of MS calculations for both trigonally and tetrahedrally coordinated boron are shown below the experimental curves in *Figure 6.4*. These calculations have employed only a single coordination shell and confirm the description of these spectral shapes in terms of a fingerprint of the local coordination. In *Figure 6.5* we show a comparison of a calculated band structure and two different MS modelling techniques for the C *K*-ELNES in titanium carbide, TiC. Here, as is more usual, the MS calculations have employed a much larger cluster of seven shells to achieve acceptable agreement with the experimental C *K*-ELNES, thus demonstrating that in this particular case the ELNES is sensitive to the medium range order in the material (Scott *et al.*, 2001).

Currently, comparison of theory and experiment is done in a mainly qualitative fashion. However, in recent years there have been considerable improvements in both the range and performance of modelling techniques, the use of parallelized software codes, as well as significant increases in computing power. One benefit of these advances has been the use of accurate approximations for the problems of electron exchange and correlation in quantum mechanical calculations of the electron wavefunctions in solids; this is almost exclusively performed using density functional theory (DFT) and the local density approximation (LDA) which assumes this exchange-correlation contribution to the overall potential in a particular region of space is a function of the local charge density. A second related benefit has allowed the routine inclusion of self-consistency in the calculations (i.e. the system is allowed to reach the lowest total energy situation via the iterative movement of electron density within the structure). Furthermore, it is now more common to include the effects of the transition process itself, notably the creation of a

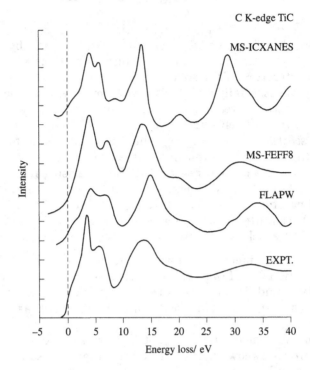

**Figure 6.5.** Comparison of the experimental C *K*-ELNES from TiC (*EXPT.*) with the results of theoretical modelling calculations using both a band structure code (*FLAPW*), and two different MS codes (*FEFF8* and *ICXANES*).

*core hole* in the inner shell during electron excitation and its effects on the different symmetry components of the DOS. Often the core hole can act to lower the energy of empty states near the edge onset and lead to an increase in the relative intensity in this region (Brydson, 1991). Hence, in the near future, a more quantitative reliability of fit should be achievable for the modelling of both peak intensities and absolute energy positions of ELNES features.

# 6.3 Determination of valencies

The valence or oxidation state of the particular atom undergoing excitation influences the ELNES in two distinct ways. Firstly, changes in the effective charge on an atom lead to shifts in the binding energies of the various electronic energy levels (both the initial core level and the final state) which often manifest in an overall *chemical shift* of the edge onset. These shifts are similar, although not identical in origin, to those commonly observed using the surface sensitive technique of XPS discussed in Section 1.4.2. For example, EELS measurements on tri- and tetravalent titanium standards show that the Ti $L_3$ peak occurs some

2 eV lower in energy for $Ti^{3+}$ as compared to $Ti^{4+}$; many other elements also exhibit such chemical shifts when incorporated in different compounds. Many of these shifts have been empirically tabulated (Brydson *et al.*, 1992a); however, currently there is no reliable means for the theoretical prediction of absolute edge onset energies. An application of this method, shown in *Figure 6.6*, is concerned with the determination of the proportion of $Ti^{3+}$ introduced by the electrochemically induced nucleation of a fresnoite glass ceramic (fresnoite normally contains solely $Ti^{4+}$ in an unusual fivefold coordination). The small downward shift in energy (relative to pure fresnoite) suggested the presence of less than 10% $Ti^{3+}$ in the glass neighbouring the electrode, which was confirmed via the observation of a corresponding increase in pre-peak intensity (*ca.* 5%) at the O $K$-edge due to the presence of charge compensating oxygen vacancies (Hoche *et al.*, 2001).

Secondly, the valence of the excited atom can affect the intensity distribution in the ELNES. This predominantly occurs in edges that exhibit a strong interaction between the core hole and the excited electron, leading to the presence of *quasi-atomic* transitions. These are so called since the observed ELNES is essentially atomic in nature and only partially modified by the crystal field due to the nearest neighbour atoms. Examples of such spectra are provided by the $L_{2,3}$-edges of the 3d and 4d transition metals and their compounds and the $M_{4,5}$-edges of the rare-earth elements. These spectra exhibit very strong, sharp features known as *white lines* which result from transitions to energetically narrow d or f bands. This makes detection and quantification of these elements extremely easy. *Figure 6.7* shows the $L_{2,3}$-edges from a variety of

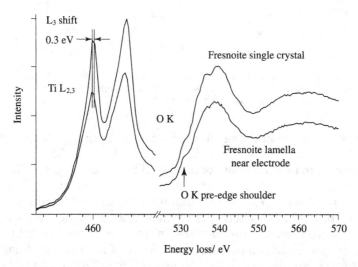

**Figure 6.6.** EELS spectra showing the Ti $L_{2,3}$- and O $K$-edges from a fresnoite single crystal and a fresnoite lamella, produced by electrochemically induced crystallization of a glass, near to the electrode. The small downward shift in the $L_3$ peak position suggests only a small reduction in the Ti valency in the electrochemically produced material consistent with a small change in the O $K$ pre-edge peak intensity indicative of O vacancies.

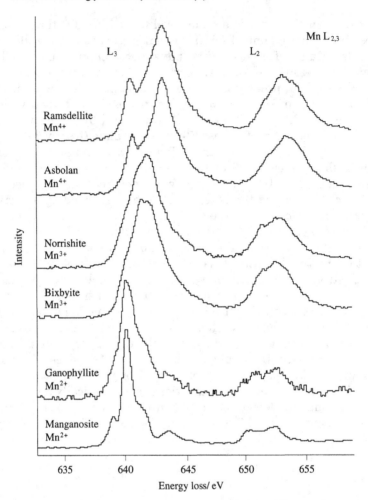

**Figure 6.7.** Comparison of Mn $L_{2,3}$-ELNES from a range of manganese minerals showing the systematic change in the $L_3$ peak position, the $L_3/L_2$ white-line intensity ratio and the detailed fine structure as a function of the oxidation state of Mn.

manganese oxides, containing manganese in different oxidation states (Garvie *et al.*, 1994). As may be seen, transitions from the Mn $2p$ shell are actually split into two components separated by the spin orbit splitting of the ionized $2p$ core level: an $L_3$-edge followed at higher energy loss by a $L_2$-edge. From *Figure 6.7*, it is clear that both the $L_3$- and $L_2$-edges exhibit a chemical shift to higher energy loss with increasing manganese valency. Additionally, the relative intensities in the two separate white-line components also clearly depend on the valence of the excited atom, thus leading to the possibility of oxidation-state identification. This dependence of the white-line intensity ratio on the valence is due to the different interactions between the two core hole initial states and the electrons in the final state. By using an appropriately normalized sum of the white-line intensities it may be possible to derive a linear relationship between the total white-line intensity and $d$-band occupancy for both the $3d$ and $4d$ transition metals as well as the $f$-band occupancy for the rare earths (Brydson *et al.*, 1992a). Closer inspection of *Figure 6.7* reveals that the detailed shape of both white-line components differs for differing oxidation states. These shapes are most conveniently modelled by *atomic multiplet theory* which allows calculation of the atomic transitions of the particular ion in the field due to the nearest neighbour atoms (de Groot *et al.*, 1990). Such spectra provide *ELNES valence state fingerprints* for transition metal and rare earth ions in various environments (e.g. octahedral, tetrahedral, square planar etc.).

Quantification of the relative proportions of differing valence states of a particular element in a material using the techniques outlined above has been demonstrated and represents an extremely exciting area in microstructural analysis particularly in the field of geochemistry (Garvie and Buseck, 1998).

## 6.4 Determination of bond lengths

Some tens of eV beyond the near-edge or ELNES region, superimposed on the gradually decreasing tail of the core loss edge, one observes a region of broader, less-intense ELNES features followed by much weaker, extended oscillations known as the *EXELFS*. In terms of scattering theory, the main distinction between the ELNES and EXELFS regions lies in the fact that the low kinetic energy of the ejected electron in the near-edge region means that it samples a greater volume before it loses its energy and is brought to a halt (i.e. the inelastic mean free path is large); thus multiple elastic scattering can occur so providing geometrical information as discussed in Section 6.2.1. Further above the edge onset, the higher kinetic energy results in a shorter inelastic mean free path and therefore predominantly single elastic scattering of the ejected electron, so giving short-range information which may be used to extract simple information on quantities such as bond lengths. This apparently curious behaviour arises since the dominant inelastic scattering process in a solid, contributing to the stopping of the excited electron produced during inner-shell ionization, is plasmon excitation which typically requires a minimum energy of between 15 and 30 eV.

There are two approaches to bond length determination. The first is to use the energy position of the broad ELNES peaks some 20–30 eV above the edge onset, known as *multiple scattering resonances (MSR)*. As their name suggests, these features arise from a resonant scattering event involving the excited electron and a particular shell of atoms. The energies of these features above the edge onset have been shown to be proportional to $1/r^2$, where $r$ is the bond length from the ionized atom.

Identification of such MSR permits a semi-quantitative determination of nearest neighbour, and in some cases second nearest neighbour bond lengths (Kurata *et al.*, 1993).

The more standard procedure is to analyse the weak EXELFS oscillations occurring some 40–50 eV above the edge onset. Since these oscillations are weak, high statistical accuracy (i.e. high count rates and long acquisition times) is required if useful information is to be extracted. EXELFS oscillations may be thought of as arising from the interference between the outgoing electron wave from the excited atom and the wave that has been backscattered (once) from the neighbouring shells of atoms. After deconvolution to remove thickness effects, the processing of EXELFS data is identical to that used in EXAFS data processing (see Section 1.4.2). This involves the extraction of the EXELFS oscillatory component via the fitting of a low-order polynomial to the decreasing edge intensity beginning some few tens of eV above the edge onset; note, to obtain good data a large energy range of typically a few hundred eV is required, which may present problems in systems possessing edges closely spaced in energy. The oscillatory data is then transformed from an energy scale to a wavevector ($k = 2\pi/\lambda$) scale, smoothed and possibly truncated. In order to determine the amplitude and phases of the oscillations, a Fourier transform is taken which, after correction for the phase shifts occuring as a result of backscattering, effectively gives a *radial distribution function* (RDF) of neighbouring atoms around the ionized atom. Further details of the EXELFS data processing procedure are given in Egerton (1996). As an example, *Figure 6.8* shows the RDFs derived from the oxygen $K$-EXELFS of α-alumina and amorphous alumina (Bourdillon *et al.*, 1984). The position of the first peak in the transform gives the Al–O distance, which exhibits a contraction in the case of the amorphous material. The areas under the various peaks correspond to the coordination numbers, although these are extremely

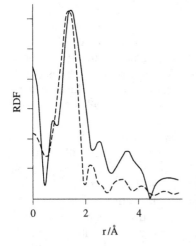

**Figure 6.8.** The RDFs obtained by Fourier transformation of the oxygen $K$-EXELFS oscillations of α-alumina (solid line) and amorphous alumina (dashed line) showing differences in Al–O bond lengths. Note the RDF peak maxima need to be corrected for the backscattering phase shifts to give direct bond lengths.

sensitive to the accurate correction of non-uniformities in the EELS detection system.

# 6.5 Experimental aspects of ELNES/EXELFS measurements

In order to investigate and characterize the modifications in the ELNES and EXELFS associated with the bonding of a particular atom, the importance of recording spectra with good, accurate statistics can not be stressed too highly. Similar care is also required for accurate elemental quantification as outlined in Chapter 5. Furthermore, good energy resolution is also paramount for detailed ELNES measurements, particularly where the measurement and modelling of fine spectral details are required; EXELFS can often be performed at lower energy resolutions due to the low-frequency nature of the extended oscillations.

As discussed in Section 3.3.2, parallel detection systems, whilst providing good detective quantum efficiency, influence the data by adding additional noise to the raw data as well as producing a non-uniform response to the input signal across the length of the detector. A number of routine corrections are therefore required prior to further signal processing. The finite detector dark count, which is the detector read-out signal in the absence of illumination, is best measured immediately prior to, and/or after, data acquisition using the same acquisition parameters. This avoids the problem of 'ghost peaks' caused by overexposure of the detector array. The non-uniform channel-to-channel gain appears as both high- and low-frequency variations in the data. The high-frequency components appear as noise on the as-recorded spectrum and correcting these variations is crucial for optimizing detection sensitivity. Uncorrected, the lower-frequency components may be misiden-tified as ELNES, but may be readily identified by displacing the spectrum across the detector array. As outlined in Section 3.3.2, two methods of gain correction are frequently employed, one involving uniform illumination of the detector to produce a gain normalization function for subsequent correction of spectra, while the second involves shifting the spectrum across the detector array by applying small voltage adjustments to the spectro-meter drift tube – separate spectra are then realigned and summed.

Section 3.5.2 discusses the factors determining the spectral energy resolution, which largely depends on the energy spread of the electron source. For cold field emission sources, a resolution of 0.3 eV is attainable; thermally assisted field emission sources routinely provide 0.8 eV, while $LaB_6$ sources result in resolutions of between 1 and 1.5 eV. Monochroma-tion of the source is also possible but not yet widely commercially available. It is important to realize that for a given spectrometer system focussed at the ZLP, the energy resolution degrades with increasing energy loss. One option for the recording of high-energy loss edges is to focus the spectrometer using a sharp ELNES feature such as a white line, although care should be exercised so as not to introduce apparent features in the measured ELNES. For a given energy resolution, ELNES measurements generally require use of the highest possible energy dispersion, typically 0.1–0.3 eV/channel, provided this does not unduly degrade statistics. Further corrections to the data which are necessary in high-resolution systems are due to the apparent degradation of the resolution due to the spreading electron beam and light in the scintillator. The system point spread function, measured by reducing the dispersion such that all the elastically scattered electrons in the ZLP are focussed to a spot smaller than the physical dimension of a single diode, reflects the spreading of the beam by the scintillator and may be used to routinely deconvolve all spectra using Fourier Transform techniques. Additionally ELNES data may be sharpened by deconvolving the measured ZLP, but this may be limited by noise considerations.

An accurate means of absolute energy calibration is also desirable if the small energy shifts arising due to changes in the charge distribution are to be measured and compared. One method is to monitor the position of an edge relative to a second edge measured in the same spectrum; adventitious carbon contamination on the specimen can be useful in this respect. Another way is to apply a known voltage to the electron drift tube of the spectrometer (or, in some machines, change the accelerating voltage of the microscope) and calibrate by reference to a known edge energy (e.g. the Ni $L_3$-edge in NiO–see Section 3.5.2). The voltage required to displace the ELNES structure of interest to the ZLP can then be used to determine edge positions to an accuracy of around $\pm 0.2$ eV. Future methods may rely on a self-compensating feedback system between the high voltage supply and the spectrometer drift tube.

From Section 2.4, inelastic scattering results in a Lorentzian angular distribution with a characteristic angle given by $\theta_E = E/2E_o$. To optimize sensitivity for a given edge, an aperture size (either objective aperture in CTEM image mode, spectrometer aperture and camera length in CTEM diffraction mode or collector aperture in STEM) should be chosen which accepts the majority of inelastically scattered electrons. However, increasing the collection angle, $\beta$, increases both the background beneath the edge and the fraction of electrons which have undergone non-dipolar transitions. It is important to collect electrons which have transferred small amounts of momenta since this ensures that the resultant ELNES are then dominated by dipole allowed transitions and ELNES spectra may then be directly compared with those obtained using X-ray absorption spectroscopy (XAS and XANES) and more easily modelled using theoretical electronic structure calculations since the final states of the excitation process are then given, in the atomic picture, by $\Delta l = \pm 1$, where $\Delta l$ is the change in angular momentum quantum number between the core level and the unoccupied final states. Overall a good compromise is to employ a collection angle of around 2–3 times $\theta_E$.

Processing of ELNES data should be performed on data that has been background subtracted. This can be difficult, especially for low-energy edges, and various strategies for background subtraction are discussed in Chapter 5. An optimum sample thickness, $t$, for detection sensitivity and signal-to-noise ratio, whilst avoiding the effects of excessive multiple inelastic scattering, is about $0.5t/\lambda$. However, even if multiple inelastic scattering effects are small, whenever possible spectra should be deconvoluted using the standard techniques described in Section 4.1, if accurate comparisons and modelling of near-edge features are to be achieved. The latter is particularly true if quantitative measurements of site occupancies are to be obtained.

Finally, if possible, ELNES measurements should be performed on polycrystalline sample areas so as to avoid orientation-dependant effects – unless, as is discussed in Chapter 8, such information is required for the assignment of near-edge features and the extraction of directional bonding information.

## 6.6 Conclusions

To conclude, the ELNES and EXELFS signals observable on an inner-shell ionization edge, while comparatively weak, can offer highly significant information for both materials chemists, such as coordinations, valencies and bond lengths associated with elemental species, and solid-state physicists, where fine details of the electronic structure are required. Analysis and modelling of EXELFS data is well established, although experimental measurement is generally difficult to perform, owing to the fact that a weak signal has to be recorded over a large energy region. ELNES data is more intense and, apart from the requirements of good energy resolution, experimentally the recording is therefore considerably more straightforward. Conversely, accurate modelling of ELNES data is extremely difficult and time consuming; as a result various empirical approaches such as fingerprinting techniques have proved both popular and successful.

Section 1.4.2 outlines the direct equivalence of ELNES and EXELFS to XANES and EXAFS measurements, which have been extremely useful methods for general structural and chemical characterization of solids and surfaces; however, the ability to perform EELS measurements at high spatial resolution in the TEM can provide important advantages for the characterization of microstructure, some of which are described in Chapter 8.

## References and further reading

**Berger, S.D., McKenzie, D.R. and Martin, P.J.** (1988) EELS analysis of vacuum arc deposited diamond-like films. *Phil. Mag. Lett.* **57:** 285.

**Bourdillon, A.J., El-Mashri, S.M. and Forty, A.J.** (1984) Application of TEM EXELFS to the study of aluminium oxide films. *Phil. Mag. A* **49:** 341.

**Brydson, R.** (1991) Interpretation of near-edge structure in the electron energy loss spectrum. *EMSA Bulletin*, Fall edition 21:2.

**Brydson, R., Sauer, H. and Engel, W.** (1992a) ELNES as an analytical tool – the study of minerals. In: *Transmission Electron Energy Loss Spectrometry in Materials Science* (eds M.M. Disko, C.C. Ahn and B. Fultz). TMS, Warrendale, Pennsylvania.

**Brydson, R., Sauer, H., Engel, W. and Hofer, F.** (1992b) Near-edge structures at the O $K$-edges of Ti(IV) oxygen compounds. *J. Phys.: Condensed Matter* **4:** 3429–3437.

**de Groot, F.M.F., Fuggle, J.C., Thole, B.T. and Sawatzky, G.A.** (1990) $2p$ X-ray absorption spectra of $3d$ transition metal compounds: an atomic multiplet description including the crystal field. *Phys. Rev. B* **42:** 5459.

**EELS and X-ray database** – *http://www.cemes.fr/~eelsdb/*

**Egerton, R.F.** (1996) *Electron Energy Loss Spectroscopy in the Electron Microscope*. Plenum Press, New York.

**Garvie, L.A.J. and Buseck, P.R.** (1998) Ratio of ferrous to ferric iron from nanometre sized areas in minerals. *Nature* **396:** 667–669.

**Garvie, L.A.J., Craven, A.J. and Brydson, R.** (1994) Use of electron loss near-edge fine structure in the study of minerals. *Am. Mineralogist* **79:** 411–425.

**Garvie, L.A.J., Craven, A.J. and Brydson, R.** (1995) Spatially resolved measurements of boron environments in ultrafine minerals: an electron energy loss spectroscopic study. *Am. Mineralogist* **80:** 1132–1144.

**Garvie, L.A.J., Rez, P., Alvarez, J.R., Buseck, P.R., Craven, A.J. and Brydson, R.** (2000) Bonding in alpha-quartz ($SiO_2$): a view of the unoccupied states. *Am. Mineralogist* **85:** 732–738.

**Hoche, T., Kleebe, H. J. and Brydson, R.** (2001) Can fresnoite, $Ba_2TiSi_2O_8$, incorporate $Ti^{3+}$ when crystallising from highly reduced melts? *Phil. Mag. A* **81:** 825.

**Hofer, F. and Golob, P.** (1988) New examples for near-edge fine structures in EELS. *Ultramicroscopy* **21:** 379.

**Kurata, H., Lefevre, E., Colliex, C. and Brydson, R.** (1993) ELNES structures in the oxygen $K$-edge spectra of transition metal oxides. *Phys. Rev. B* **47:** 13763–13768.

**Rez, P.** (1992) Energy loss fine structure. In: *Transmission Electron Energy Loss Spectrometry in Materials Science* (eds M.M. Disko, C.C. Ahn and B. Fultz). TMS, Warrendale, Pennsylvania.

**Sauer, H., Brydson, R., Rowley, P., Engel, W., Thomas, J.M. and Zeitler, E.** (1993) Determination of coordinations and coordination-specific site occupancies by electron energy loss spectroscopy: an investigation of boron–oxygen compounds. *Ultramicroscopy* **49:** 198–209.

**Scott, A.J., Brydson, R., MacKenzie, M. and Craven, A.J.** (2001) A theoretical investigation of the ELNES of transition metal carbides and nitrides for the extraction of structural and bonding information. *Phys. Rev. B.* **63:** 245105.

# 7   EELS imaging

## 7.1  Introduction to EELS imaging and energy filtering

In recent years, associated particularly with the development of two-dimensional, parallel detection systems, there has been increasing interest in exploiting the potential imaging capabilities of EELS. Using electrons of a specific energy loss or range of energy losses to form images of specimen thickness, elemental composition, phase composition or even electronic structure represents an extremely attractive and exciting prospect for the microstructural analyst, particularly due to the inherent high spatial resolution of the analytical technique detailed in Section 5.7.

One approach to EELS imaging, in widespread use for the characterization of both biological and materials science samples in the TEM, is to employ a magnetic prism spectrometer and an energy selection slit to filter the spectrum of transmitted electron energies normally used to form conventional TEM images and diffraction patterns. Only electrons of a particular energy loss are transmitted so forming an energy-filtered image or diffraction pattern. This is most commonly known as either *EFTEM, ESI* or the *'imaging spectrum'* technique. Using only elastically scattered electrons (i.e. those contained in the ZLP) to form images and diffraction patterns increases contrast and resolution, allowing easier interpretation than with unfiltered data. Furthermore, chemical mapping may be achieved by acquiring and processing images formed by electrons that have undergone inner-shell ionization events. An alternative approach, used when mapping is performed with a PEELS attached to a STEM, is to raster the electron beam across the specimen and record an EELS spectrum at every point $(x,y)$ – this technique being known as *'spectrum imaging'*. The complete three-dimensional dataset $(x,y$ and energy loss) may then be subsequently processed to form a two-dimensional quantitative map as a function of position on the sample area using either standard elemental quantification procedures, or the position and/or intensity of characteristic low loss or ELNES features.

In comparison with spectrum imaging, the ESI approach, on which we will concentrate (at least initially), involves collecting a two-dimensional image with one energy window, different energies having to be explored sequentially. This is achieved by passing the demagnified image through the spectrometer, selecting the electrons of a particular energy loss range using a slit of variable width, and then remagnifying the image. Multiple images may then be processed to form, say, an elemental distribution or concentration map, an example of which is shown in *Figure 7.1*. Some dedicated energy filtering microscopes (EFTEM) employ a Castaing–Henry filter or $\Omega$ filter situated within the microscope column (shown in *Figure 3.5*; Egerton, 1996; Reimer, 1995), whereas conventional TEMs use a post-column imaging spectrometer (commercially known as a GIF) which is essentially an 'add-on' attachment to the bottom of the microscope column. The latter is obviously a more flexible approach

**Figure 7.1.** (a) TEM bright field image, together with: (b) Ti $L_{2,3}$- and (c) Nb $M_{4,5}$-edge EFTEM elemental maps of precipitates extracted from a triple microalloyed steel after controlled rolling. Note the presence of niobium carbide caps on cuboid titanium nitride precipitates. (d) High-resolution image of (002) lattice planes at the coherent interface between TiN and NbC cap showing epitaxial growth.

and is shown in *Figure 3.4.* A single magnetic sector spectrometer together with alignment and focusing quadrupoles (essentially as in a normal PEELS system) is followed by an energy selecting slit and a number of quadrupole and sextupole lenses. The image is recorded on a two-dimensional slow-scan CCD camera. The quadrupoles after the selecting slit can either project a focused image of the spectrum formed at the slit (spectrum mode), or they may cancel the energy dispersion produced by the spectrometer and project a magnified image of the specimen (image mode). In the post-column GIF, the magnification on the CCD relative to the TEM viewing screen is approximately 20 times; the inherent magnification of the post-column filter means that EFTEM studies are normally performed at TEM magnifications of between 1000 and 40 000 times. To record a conventional EELS spectrum, the energy selecting slit is removed and the filter operated in spectrum mode, the spectrum incident on the detector is spread out in the non-dispersion direction and spectrum profiles obtained by lateral integration over the detector pixels.

## 7.2 Summary of energy-filtering techniques

One of the first extensive applications of EFTEM has been in the examination of biological specimens, which may be thick and are generally of low atomic number, thus scattering weakly. Furthermore, owing to their beam sensitivity, TEM of biological materials tends to be performed at relatively low accelerating voltages so as to avoid knock-on damage. All these contributing factors generally result in the production of very low contrast images. One early solution to this problem was the conventional staining of thin sections with high atomic number elements such as osmium. However, in recent decades, an alternative solution has been EFTEM *contrast tuning*, where image contrast is enhanced by selecting a particular range of energy losses such as the ZLP, the most probable energy loss in thick samples (typically 30–40 eV), or the pre-ionization edge background of the most predominant element, carbon (Egerton, 1996). More generally, in materials science applications, both EFTEM and spectrum imaging techniques may be most conveniently categorized via consideration of the specific region of the EEL spectrum, which is transmitted to form the resultant image (or diffraction pattern in the case of EFTEM).

(i) *ZL Filtering* – TEM bright field images and diffraction patterns contain an inelastically scattered component, which increases with increasing sample thickness and is usually assumed to produce a diffuse background contribution. If a narrow (*ca.* 5 eV) transmitted energy window is placed over the ZLP, this filters out the majority of this inelastic component and may significantly improve image contrast or reveal weak spots and/or structure in a diffraction pattern. Although ZL filtering reduces overall image intensity, image resolution is improved, particularly for thicker specimens, due to the reduction in chromatic aberration. In the case of high-resolution phase contrast or lattice imaging, accurate image interpretation and simulation using multi-slice techniques may only be possible using ZL filtered images. ZL filtering has been used to determine radial distribution functions from diffraction patterns of amorphous materials and is also used extensively to enhance convergent beam electron diffraction patterns and derive detailed information on lattice parameters and bonding effects (Midgley *et al.*, 1995).

(ii)    *Thickness mapping* – if both an unfiltered image and a ZL filtered image are recorded from the same specimen area and the intensity at each pixel processed according to Equation (4.1) (i.e. the two image intensities are divided and the logarithm taken), then the result is essentially a map of specimen thickness normalized by the inelastic mean free path – also known as a $t/\Lambda$ map. The values of $t/\Lambda$ can be converted to absolute thickness if $\Lambda$ is known; however, in a heterogeneous specimen $\Lambda$ will vary between different phases. As we shall see such thickness maps can be used to correct and derive absolute concentrations from elemental maps.

(iii)   *Low loss imaging* – the position of the plasmon peak is sensitive to the valence electron density (Equation 4.3), hence positioning a narrow (*ca.* 1–2 eV) energy selecting window in the low loss region of the EEL spectrum can allow the mapping of distinct alloy compositions or phases within a microstructure via the direct imaging of shifts in the plasmon peak energy. An example of a plasmon map produced by the spectrum imaging technique is shown in *Figure 4.4*. Further possibilities include the imaging of low-energy surface plasmon features or interband transitions characteristic of a given phase within a microstructure (Chapter 4).

(iv)    *Elemental mapping* – elemental maps can be formed by imaging with electrons which have lost energies corresponding to the characteristic inner-shell ionization edges employed for quantitative point analysis described in Chapter 5. This important technique, discussed in detail in Section 7.3, typically employs energy windows of between 10 and 100 eV. *Figure 7.1* shows an example of a set of elemental maps of nanoscale precipitates extracted from a microalloyed steel. These maps clearly identify the presence of niobium-rich caps on titanium nitride precipitates, which arise from the kinetics of precipitation in the steel during weld thermal cycles (Egner *et al.*, 1999).

(v)     *Chemical state mapping* – as we have seen in Chapter 6, the ELNES associated with a particular inner-shell ionization edge may contain features which are characteristic of a particular type of coordination or valence state of an elemental species. The use of a narrow energy window positioned on such an ELNES signal, e.g. the $\pi^*$ peak associated with the presence of $sp^2$-bonding (*Figure 6.4*), allows the mapping of the variation in intensity in this feature as a function of position within the microstructure. This will produce a chemical state map for a particular elemental species. Examples of this technique have been demonstrated for mapping $sp^2$-bonded carbon (Muller *et al.*, 1993) and also the valency of transition metal ions (Wang *et al.*, 2000).

Although energy filtering is routinely used to remove the inelastically scattered electron component in conventional imaging or diffraction studies, elemental mapping remains one of the most promising applications of the technique. Elemental mapping using PEELS can be carried out in both scanning and fixed-beam (conventional) TEM and a comparison of these two techniques has been extensively discussed (Reimer, 1995) and is summarized in Section 7.8.

In the STEM, *spectrum imaging* data consists of an individual PEEL spectrum recorded at every probe position which may be quantified by post-acquisition processing using the procedures outlined for the quantification of PEEL spectra in Chapters 4 and 5. The results may then be subsequently redisplayed as a two- or three-dimensional map of concentration or relative concentration as a function of probe position. In comparison, in the CTEM, *imaging spectrum* or *ESI* data consists of a map of image intensity as a function of energy loss. In principle, for amorphous and biological samples, the image intensity at any pixel in an EFTEM image can be converted to a mass or concentration using directly analogous procedures to those applied in spectral quantification; however, for crystalline materials, the electron scattering no longer follows a simple atomic form but also includes significant diffraction effects, such as bend and thickness contours. The latter contributes significantly to EFTEM images and makes quantification considerably more difficult (Hofer and Warblicher, 2001).

# 7.3 Procedure for EFTEM elemental mapping

Quantification of EFTEM images recorded using an energy loss window, of width $\Delta$, positioned on the ionization edge of interest (a *post-edge image*) requires the following steps (Hofer *et al.*, 1997; Hofer and Warblicher, 2001).

(i)    Correction of all images for dark current, detector response (gain) and point spread function of the scintillator. In an analogous manner to PEEL spectroscopy, the dark current contribution is measured by blanking the beam and integrating the signal at each pixel of the CCD over an identical time period as that used to record the post- and pre-edge images.

(ii)    Removal of background contribution to the post-edge image intensity. In most cases a power law background is assumed and the so-called *three-window method* is employed. Here, two EFTEM images, also recorded with an energy window of width Δ, now positioned in front of the edge (*pre-edge images*), as well as the post-edge image are recorded and each are corrected for dark current, gain and PSF. The two pre-edge images are used to produce an extrapolated background image, which is calculated via the two-area method (Section 5.2). This extrapolated background image is then subtracted from the post-edge image to produce an *elemental map*, as in *Figure 7.1*, where image intensity is directly related to areal density. A schematic diagram of the position of the various energy windows is shown in *Figure 7.2*. A varient on the three-window method using an extra fourth window, has been developed for closely spaced or overlapping edges (Bentley, 1998).

(iii)    Since post-edge and pre-edge images are recorded sequentially, sample drift may occur between the recording of these successive energy-filtered images. Generally specimen drift during EFTEM measurements may be minimized by use of short integration times, although this is obviously at the expense of increased noise in the images. Before any processing of images, for example to remove the post-edge background contribution, it is therefore important that all images are spatially registered with respect to each other via the use of automated cross-correlation algorithms and/or careful manual alignment. Such cross-correlation is greatly aided by the presence of sharp features in the images, such as the edge of the

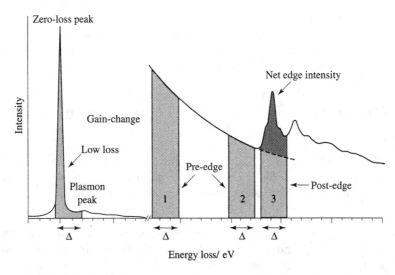

**Figure 7.2.** Typical EEL spectrum showing the low loss region (ZL and plasmon peak) and an inner-shell ionization edge: the positions of the energy windows required to form energy-filtered images for background subtraction and quantification are indicated by the shaded areas.

specimen, or a phase boundary. Furthermore for accurate correlation the images should all be in focus. From Section 3.4, in order to
minimize chromatic aberration effects, offsetting the spectrum
across the energy selecting slit in EFTEM is achieved by changing
the microscope accelerating voltage rather than the spectrometer
drift tube voltage. This change will generally lead to image defocus
(unless the microscope is accurately voltage centred). Hence,
wherever possible post-edge images should be refocused before
recording.

(iv) Relative quantification of elemental maps can be achieved by
normalization of the image intensity, $I$, by the appropriate partial
ionization cross-section for an energy window $\Delta$ and the relevant
experimental conditions $E_0$, $\alpha$ and $\beta$. The product of atomic
concentration, $n$, and thickness, $t$ (the areal density), is given by:

$$n.t(\underline{x},\underline{y}) = I(\underline{x},\ \underline{y},\ \Delta,\ \underline{E_0},\ \underline{\alpha},\ \underline{\beta})/\sigma(\Delta,\ \underline{E_0},\ \underline{\alpha},\ \underline{\beta}) \tag{7.1}$$

After cross-correlation, two relative concentration elemental maps
may then be divided to produce an *atomic ratio map* in which
effects due to variations in both sample thickness and diffraction
contrast can often cancel, especially if the two ionization edges lie
within the same energy loss range. This technique is highly
applicable to polycrystalline materials science samples.

$$\frac{n_A(\underline{x},\underline{y})}{n_B(\underline{x},\underline{y})} = \frac{[I_A(\underline{x},\ \underline{y},\ \Delta,\ \underline{E_0},\ \underline{\alpha},\ \underline{\beta})/\sigma_A(\Delta,\ \underline{E_0},\ \underline{\alpha},\ \underline{\beta})]}{[I_B(\underline{x},\ \underline{y},\ \Delta,\ \underline{E_0},\ \underline{\alpha},\ \underline{\beta})/\sigma_B(\Delta,\ \underline{E_0},\ \underline{\alpha},\ \underline{\beta})]} \tag{7.2}$$

(v) Absolute quantification requires elemental maps to be divided by
both the partial ionization cross-section and also the corresponding
low loss image, i.e. an image obtained with an identical energy
window, $\Delta$, containing the ZL and plasmon peak(s) – see *Figure 7.2*.

$$n.t(\underline{x},\underline{y}) = I(\underline{x},\ \underline{y},\ \Delta,\ \underline{E_0},\ \underline{\alpha},\ \underline{\beta})/[\sigma(\Delta,\ \underline{E_0},\ \underline{\alpha},\ \underline{\beta}).I_{LL}(\underline{x},\ \underline{y},\ \Delta,\ \underline{E_0},\ \underline{\alpha},\ \underline{\beta})] \tag{7.3}$$

Equation (7.3) demonstrates that the low loss image needs to be
recorded under identical experimental conditions ($\underline{\Delta}$, $\underline{E_0}$, $\underline{\alpha}$, $\underline{\beta}$) if
mass-thickness effects are to be accurately removed by the division
of image intensities. Owing to the large differences in image
intensity, this necessitates the adjustment of solely the beam
current and the integration time between recording the elemental
map and the low loss image. Again, before this division, the images
need to be cross-correlated. Note that this correction procedure
provides images where the intensity is equal to areal density and
not concentration (Section 5.4). Variations in sample thickness can
be partially corrected for by dividing the areal density map
produced above by a thickness map; this results in a true
concentration map. The correction for thickness is only partial
since an average mean free path for the whole specimen is usually
assumed. One major problem in these quantitative maps is

residual intensity variations due to diffraction effects in crystalline materials; these arise since diffraction effects can be significantly different in images recorded at different energy losses and therefore do not necessarily disappear during the division of core loss and low loss images (Hofer *et al.*, 1997; Hofer and Warblicher, 2001). Using the TEM techniques of rocking beam (Hofer and Warblicher, 1996) or hollow cone (Marien *et al.*, 1999) illumination, it is possible to vary the direction of the incident beam during the recording of the energy-filtered images. This causes local diffraction contrast in images to be smoothed and diffraction-induced artefacts in processed elemental maps are minimized.

If the element of interest is present at low levels, the *three-window method* often produces noisy images and artefacts in elemental maps. An alternative, more rapid procedure is simply to divide the post-edge image by a pre-edge image with the energy window positioned just before the edge onset. The result is a so-called *jump-ratio image*. This technique tends to reduce added noise in maps of elemental distributions and such jump-ratio images are also found to be significantly less affected by diffraction effects, which are invariably present in crystalline materials. Although small variations in sample thickness are often accounted for, jump-ratio images are susceptible to large changes in the slope of the pre-edge background, which cause image artefacts due to variations in sample thickness. Despite being relatively fast, simple to acquire and readily interpretable, it is important to appreciate that these jump-ratio images are inherently *non-quantitative*.

*Figure 7.3* shows a comparison of an elemental map and a jump-ratio image.

## 7.4 Experimental parameters in EFTEM

The two fundamental parameters in an EFTEM *elemental* map are: (i) the detectability limit and (ii) the spatial resolution in the image. In practice, both these factors are largely determined by the SNR in the image. Experimental parameters for optimizing the signal to noise ratio in a variety of elemental maps have been discussed extensively by Berger and Kohl (1993). Some of these parameters are general microscope operating parameters, whereas others depend on the ionization edge in question. Often, the choice of certain parameters involves making a compromise between the requirements for low detection limits and high resolution. In most applications, it is usually the limit of detectability which is the most important factor; however, in recent years some extremely high-resolution EFTEM maps have been demonstrated which open up the possibility for near-atomic resolution EFTEM (Freitag and Mader, 1999).

**Figure 7.3.** (a) TEM bright field image, together with: (b) Al $K$-edge elemental map, (c) Fe $L_{2,3}$-edge elemental map and, (d) Fe $L_{2,3}$-edge jump-ratio image of an Fe–alumina ceramic matrix composite. Note the absence of diffraction contrast due to dislocations and bend contours in the Fe jump-ratio image as compared to the Fe elemental map.

Fundamentally, the spatial resolution of an energy-filtered image, $d$, is determined by a number of factors including: the delocalization of inelastic scattering, $R$, the diffraction limit due to the objective aperture, and the chromatic and spherical aberrations in the objective lens (Krivanek *et al.*, 1995). These contributions are summed in quadrature:

$$d^2 = R^2 + (0.6\lambda/\beta)^2 + [C_c\beta(\Delta/E_0)]^2 + (2C_s\beta^3)^2 \qquad (7.4)$$

Here $\lambda$ and $E_0$ are the incident electron wavelength and energy, $\beta$ is the collection angle as defined by the objective aperture, $\Delta$ is width of the energy window and $C_c$ and $C_s$ are the chromatic and spherical aberration coefficients of the objective lens. In Equation (7.4) the first three terms are usually most significant in determining the resolution, furthermore, the second term due to the diffraction limit is only significant for larger collection angles. The chromatic aberration term is minimized by using small energy windows and small collection angles. The inelastic delocalization limit, $R$, is an inverse function of the energy loss, and is given approximately by a function known as the classical impact parameter, $b$:

$$R = 2b = 2(h/2\pi)v/E \qquad (7.5)$$

where $v$ is the incident electron velocity and $E$ is the energy loss. More exact expressions for this function also exist which can be used in Equation (7.4) (Krivanek *et al.*, 1995).

Typical attainable spatial resolutions for EFTEM images lie in the range roughly 0.4 to a few nanometres depending on exact experimental parameters. In low loss regime, the increased delocalization limit for inelastic scattering is the primary resolution limitation; however, imaging resolutions of typically 1–2 nm have been demonstrated.

Detectability limits in EFTEM are generally much worse than for EELS spectroscopy (Section 5.7) and hence the spectrum imaging technique. This is because the usual procedure is to record only two or three images with relatively small energy windows – a limited dataset. Typically the minimum detectable mass in EFTEM is around a few hundred atoms; boundary films of approximately one monolayer have also been detected as shown in *Figure 7.4*.

Experimentally the operating parameters which affect the SNR and hence the detectability limits and spatial resolution may be summarized as follows:

(i)   The *specimen thickness* – as for general quantitative analysis, except for very thin samples the image SNR increases with decreasing specimen thickness. Thus for quantitative imaging at low detection limits, thin specimen regions (preferably of relatively uniform thickness) should be chosen and the highest possible *incident beam energy* employed.

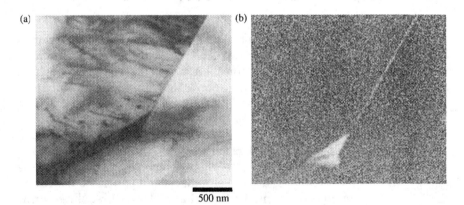

500 nm

**Figure 7.4.** (a) TEM bright field image of a liquid phase sintered alumina with additions of barium silicate, (b) the Ba $M_{4,5}$-elemental map of the same sample area showing the presence of a Ba-rich intergranular glassy film and triple pocket; from HREM and analytical STEM the intergranular film was approximately 2 nm in width and contained roughly one monolayer of Ba (thanks to Dr J. Plitzko, MPI Stuttgart).

(ii) The *current density at the specimen* – increasing the beam current and using full-cone (i.e. focussed) illumination with the largest condenser aperture all increase the image SNR. In recent years, the use of high brightness sources such as Schottky emitters has been demonstrated to work well in combination with post-column filters, which possess an inherent image magnification.

(iii) The *choice of ionization edge* – due to the rapid decrease in intensity with increasing energy loss, elemental mapping using a low-energy edge will, in principle, provide an enhanced SNR in EFTEM images. However, in thicker sample regions accurate removal of the background contribution from low-energy edges using the three-window method may make quantitative elemental mapping difficult and lead to non-physical results. Furthermore, many lower-energy edges of heavier elements exhibit a delayed edge shape (*Figure 6.1*) which can again cause problems with the removal of the background under the post-edge image intensity; where possible the choice of $K$-edges or white-line $L_{2,3}$- and $M_{4,5}$-edges are preferable in this respect. An additional problem in the low-energy regime may be the presence of overlapping edges from other elements, which may possess a different distribution within the microstructure and cause erroneous results. Under favourable conditions, high brightness sources such as field emission guns can allow mapping with edges lying in the energy range up to 3 keV. Higher-energy edges have the additional benefit of a decreased delocalization parameter, which improves the ultimate spatial resolution (Equation 7.4).

(iv) The *width of post- and pre-edge energy windows*, $\Delta$ – increasing the size of the energy selecting window will increase the image SNR, however, this will be at the expense of image resolution since chromatic aberration effects in Equation (7.4) will correspondingly increase; generally, the loss in resolution will only be apparent for high magnification energy filtering. For $K$-edges and particularly white-line $L_{2,3}$- and $M_{4,5}$-edges the majority of the edge intensity is contained in a narrow energy region near the edge onset which often allows the use of narrow energy windows whilst preserving an adequate image SNR. More delayed edges require a broader energy window, set back from the edge threshold. Larger pre-edge windows will obviously improve the reliability of the background fit, however, when using large energy windows, care should be taken that they do not include contributions from neighbouring edges both in the post- or pre-edge regions. Typical values of $\Delta$ for high-energy inner-shell ionization edges range from between 10 and 30 eV for edges below 1 keV to between 50 and 100 eV for higher-energy loss edges.

(v) The *position of post-edge energy window* relative to the edge onset depends on the basic edge shape (see *Figure 6.1*). For instance, $K$-edges and white-line $L_{2,3}$- and $M_{4,5}$-edges require an edge positioned at the edge threshold, whilst a delayed edge requires the energy window to be set back from edge threshold in order to achieve an adequate image SNR, although this may enhance errors in the removal of the background contribution.

(vi)   The *position of pre-edge energy window(s)* – generally it is advisable to position one of the pre-edge windows as close as possible to the edge threshold for the accurate removal of the background contribution to the post-edge image. Generally as the interval between the pre-edge windows increases, then so does the reliability of the background fit.

(vii)   The *value of the collection semi-angle*, β – as the microscope is in image mode this is defined by the size of the objective aperture. As discussed in Section 3.5.2, a low value of β reduces the SNR, whereas it increases the signal to background ratio making background removal more accurate. From Equation (7.4), decreasing the value of β enhances the ultimate spatial resolution but this will be at the expense of the SNR. As a general rule, β should be larger than the characteristic scattering angle, $\theta_E$, to collect a sufficiently large fraction of the inelastically scattered electrons. Thus, at 200-keV incident beam energies and for edges below about 1 keV, it is wise to use a small objective aperture (20–40 μm), whereas for higher-energy loss edges a larger objective aperture (100 μm), will provide an adequate SNR.

(viii)   The *CCD integration time* – this will depend on the signal level, which will be a function of the source brightness, the edge energy, the size of the energy window and collection semi-angle. As with all parallel detection systems, the SNR in the image depends on the optimum use of the dynamic range of the CCD and the integration time should be chosen so that the maximum signal input level is just below the saturation level of the detector. Typical integration times may vary from just a few seconds to as high as 60 s. Longer integration times may cause problems with correlation of successive energy-filtered images due to specimen and/or stage drift.

(ix)   The *microscope magnification* – for a high detection limit, the magnification should be adjusted so that the signal from the feature of interest (the specimen pixel size) does not exceed the pixel size of the detector. For instance, the optimum detectability limit for low levels of a particular element at a grain boundary, aligned edge-on with respect to the incident beam, is when the signal falls within one row of detector pixels on the CCD. Increasing the image magnification to improve image resolution would spread the signal over a number of pixels and lower the SNR. Thus detectability and spatial resolution in EFTEM images cannot be simultaneously optimized.

(x)   The *mode of operation* – for most EFTEM studies an adequate image SNR may be obtained with a reasonably low incident-beam convergence semi-angle. Alternative modes of operation include so called 'rocking beam' mode or 'hollow cone illumination' whereby a range of incident beam directions are averaged during the formation of the EFTEM images, often removing the problems of residual diffraction contrast in elemental maps.

## 7.5 Correlation of elemental maps

If a material contains a number of elements, correlation of the individual elemental maps should be performed to obtain unequivocal information on the chemical phase distributions within the sample. As already stated, since sample drift may occur between the recording of successive elemental maps, prior to image correlation, it is important that all images are spatially registered with respect to each other via the use of automated cross-correlation algorithms and/or careful manual alignment.

Correlation of elemental maps may be performed using two approaches:

(i)  The simple superposition of elemental maps to form a false colour image (see front cover image). This can be very useful for a correlation of up to three elemental maps in a red/green/blue (RGB) image, where each map is represented by a colour (R, G or B). However, colour alone is not sufficient to ascribe quantitative information to image intensity values.

(ii) The more sophisticated use of multivariate statistical tools, such as principal component analysis and multivariate histograms (scatter diagrams; Hofer *et al.*, 1997; Hofer and Warblicher, 2001). For example, to correlate two elemental maps a two-dimensional scatter diagram may be produced where the frequency of occurrence of intensity values in one elemental map is plotted against those in the second related map. Such an approach may also be extended to three dimensions. Clusters of datapoints in such scatter diagrams represent distinct chemical phases and also give information on the noise distribution and the presence of correlations or anti-correlations between elements. Such clusters may be digitally masked in the scatter plot and traceback routines allow the identification of pixel points in the original image corresponding to the clusters identified in the scatter plots.

## 7.6 General strategy for EFTEM elemental analysis

As a general rule, it is unwise to attempt EFTEM elemental mapping on a microstructure without having first confirmed that it is possible to detect the elements of interest in EELS spectroscopy mode. This is because variations is thickness and diffraction contrast can produce 'apparent maps', albeit at low contrast, even when elements are not detectable! Thus the following procedure is recommended.

(i)     Record EELS spectra from the general area of interest and determine which elements are present and detectable. Also determine the average specimen thickness from the low loss data (Equation 4.1).

(ii)    As jump-ratio images only require short exposure times, record a (two-window) jump-ratio image for each element of interest, varying the experimental acquisition parameters (Section 7.4) so as to optimize the various images.

(iii)   Using these optimized parameters, record (three-window) elemental maps for each element of interest. Process various elemental maps to obtain either atomic ratio maps or absolute concentration maps (which will additionally require the recording of a low loss image and a thickness map). Jump-ratio images can also calculated.

(iv)   Finally use the elemental maps to inform the choice of location for quantitative spectroscopy using a small probe.

## 7.7 Experimental procedure for EFTEM image acquisition and processing

Besides the general strategy outlined in the Section 7.6, the following list provides a useful summary of acquisition procedures for EFTEM imaging:

(i)      Select specimen region for analysis and microscope image magnification.

(ii)     Ensure that sample drift is low.

(iii)    Tilt to minimize diffraction contrast in image or alternatively use rocking beam or hollow cone illumination.

(iv)    Adjust image intensity using condenser lenses.

(v)     Focus and record bright field or ZL filtered image.

(vi)    Determine variation in thickness across specimen area by recording a thickness map.

(vii)   Use spectrum (i.e. high voltage) offset to select the post-edge energy loss region (or alternatively if image intensity is low use a region at say 100 eV).

(viii) Insert energy slit of width say 10–50 eV.

(ix)    Centre image intensity and re-focus image. Re-focussing will improve cross-correlation of post- and pre-edge images.

(x)     Insert objective aperture to select collection angle.

(xi)    Select onset and width of energy window for post-edge region.

(xii)   Select onset for pre-edge region 1 (and pre-edge region 2 if required).
(xiii)  Select integration time.
(xiv)   Record post-edge image.
(xv)    Record pre-edge image 1.
(xvi)   If recording an elemental map repeat for pre-edge image 2.
(xvii)  Correct all images for detector dark current, gain response and PSF.
(xviii) Correlate images.
(xix)   Process images.
(xx)    Record low loss image and/or thickness map.
(xxi)   Select regions for EELS point analysis.

## 7.8  Comparison of EFTEM and spectrum imaging methods

As stated earlier true spectrum imaging consists of sequentially scanning a small probe in a dedicated STEM or hybrid TEM/STEM and recording a complete EELS spectrum at each specimen pixel to form either a one-dimensional linescan (i.e. $x$ vs $E$) or a two-dimensional image (i.e. $(x,y)$ vs $E$).

Although only one exposure is needed for acquiring the whole map, the spectrum imaging approach is rather slow; typically 100 min are required for a $512 \times 512$ map. Another disadvantage is that large amounts of data are generated, typically 1000 Mbytes of computer memory for a $512 \times 512$ map. A further problem arises from the large dynamic range of the EEL spectrum, which may necessitate successive acquisition of a number of spectral regions at each probe position if the complete spectrum is required; this again lengthens the recording process. To circumvent this, for many applications only a part of the EELS spectrum is recorded at each probe position, for example, the high loss region if elemental mapping is desired. However, at the time of writing, recent improvements in the dynamic range and minimum integration and read-out speeds of two-dimensional CCD detectors have considerably enhanced the prospects for spectrum imaging becoming a routine tool in TEM/STEM.

Although spectrum imaging generates a large amount of data, this dataset is essentially a complete record of the analysis area and contains a great deal of detailed information. Indeed, spectrum imaging allows considerably more complex data analysis than in the case of EFTEM. Post-acquisition processing of the spectra allows the routine deconvolution of spectra to produce single scattering distributions, data may be fitted to obtain plasmon or ELNES peak positions or intensities as a function of probe position (Sections 4.2, 6.2 and 6.3), spectra may be quantitatively analysed using multivariate statistical procedures (Chapter 8) or even transformed to provide radial distribution functions

(Section 6.4) or real and imaginary parts of the dielectric function (Section 4.2). With increases in computing power and parallel processing capabilities, many procedures can now be performed on-line during acquisition. Furthermore it is easy to revisit the complete dataset and reanalyse, if new analytical questions come to light. Thus it is predicted that the importance of spectrum imaging will increase in the forthcoming years.

To obtain a more complete $(x, y, E)$ dataset using EFTEM requires a whole series of images to be recorded sequentially at differing energy losses. EELS spectra can then be reconstructed for each specimen pixel by suitable processing of the stack of images and the individual spectra then analysed or processed accordingly. Although directly complementary to the spectrum imaging technique, the major problem is that, in order to achieve reasonable acquisition times, the EFTEM image series has to be collected using energy windows of at least a few eV which can severely limit the energy resolution in the extracted EEL spectra.

One of the major benefits of spectrum imaging is that the image resolution is principally determined by the incident electron probe size and associated beam broadening in the specimen. As probe sizes continue to decrease in the new generation of aberration corrected STEMs (Krivanek *et al.*, 1997), spectrum imaging will play an important role in routine microanalysis. In terms of the analysis time for mapping elemental distributions in a complex microstructure, spectrum imaging is suited to situations where information on the distributions of a large number of elements is required, while EFTEM is considerably more efficient when information on only a few species needs to be extracted. The main disadvantage of spectrum imaging occurs when the major radiation-damage mechanism in the specimen is dose-rate dependent, here the intense focussed probe can cause local heating and degradation of the sample during measurement.

# 7.9 Energy-filtered tomography

Ultimately, the major drawback associated with any TEM-based study is the fact that all images and analysis are essentially a projected two-dimensional representation of, what is in reality, a three-dimensional specimen. Recently, a number of groups are beginning to apply a variety of methods and reconstruction algorithms, originally developed in biological imaging, to both high-resolution images and EFTEM elemental maps. Essentially this involves recording a series of EFTEM maps of elemental distributions, as a function of sample tilt and reconstructing the set of images to form a three-dimensional map (Grimm *et al.*, 1997). It is expected that these large-tilt tomographic EFTEM techniques will become increasingly important in the future.

# 7.10 Conclusions

EELS imaging may be achieved using either scanning beam techniques (spectrum imaging) or in a conventional TEM using an energy-selecting slit (energy filtering). In this chapter we have principally been concerned with energy-filtering TEM although comparisons have been drawn with the spectrum imaging approach. We have summarized the various approaches for the selection of specific regions of the EEL spectrum for the purposes of imaging and highlighted the applications of each technique. In particular, the mapping of elemental distributions in microstructures has been discussed in detail including the quantification of the intensity values in such images. Techniques of future importance, such as chemical state mapping and tomographic measurements have also been introduced.

# References and further reading

**Bentley, J.** (1998) Interfacial segregation and concentration profiles by EFTEM: issues and guidelines. *Microsc. Microanal.* **4**: 158–159.

**Berger, A. and Kohl, H.** (1993) Optimum imaging parameters for elemental mapping in an energy filtering TEM. *Optik* **92**: 175–193.

**Egerton, R.F.** (1996) *Electron Energy Loss Spectroscopy in the Electron Microscope*. Plenum Press, New York.

**Egner, D., Brydson, R. and Cochrane, R.C.** (1999) Dissolution behaviour of precipitates in the HAZ of microalloyed steels. *Inst. Phys Conf. Ser.* **161**: 447–450.

**Freitag, B. and Mader, W.** (1999) Electron specific imaging with high lateral resolution: an experimnetal study on layer structures. *J. Microsc.* **194**: 42–57.

**Grimm, R., Barman, M., Hackl, W., Typke, D., Sackmann, E. and Baumeister, W.** (1997) Energy filtered electron tomography of ice-embedded actin and vesicles. *Biophys. J.* **72**: 482.

**Hofer, F. and Warblicher, P.** (1996) Improved imaging of secondary phases in solids by EFTEM. *Ultramicroscopy* **63**: 21–25.

**Hofer, F. and Warblicher, P.** (2001) Elemental mapping using energy filltered imaging. In: *Transmission EELS in Materials Science* (eds M.M. Disko, B. Fultz and C.C. Ahn). Wiley, New York.

**Hofer, F., Grogger, W., Kothleitner, G. and Warblicher, P.** (1997) Quantitative analysis of EFTEM elemental distribution images. *Ultramicroscopy* **67**: 83–103.

**Krivanek, O.L., Dellby, N., Spence, A.J., Camps, R.A. and Brown, L.M.** (1997) Aberration correction in the STEM. *Inst. Phys. Conf. Ser.* **153**: 35.

**Krivanek, O.L., Kundemann, M.K. and Kimoto, K.** (1995) Spatial resolution in EFTEM elemental maps. *J. Microsc.* **180**: 277–287.

**Marien, J., Plitzko, J.M., Spolenak, R., Keller, R.M. and Mayer, J.** (1999) Quantitative electron spectroscopic imaging studies of microelectronic metallization layers. *J. Microsc.* **194**: 71–78.

**Midgley, P.A., Saunders, M., Vincent, R. and Steeds, J.W.** (1995) Energy filtered convergent beam electron diffraction: examples and future prospects. *Ultramicroscopy* **59**: 1–13.

**Muller, D.A., Tzou, Y., Raj, R. and Silcox, J.** (1993) Mapping $sp^2$ and $sp^3$ states of carbon at sub-nanometre spatial resolution. *Nature* **366**: 725–727.

**Reimer, L. (ed.)** (1995) *Energy Filtering Transmission Electron Microscopy*. Springer, Heidelberg.

**Wang, Z.L., Bentley, J. and Evans, N.D.** (2000) Valence state mapping of cobalt and manganese using near-edge fine structures. *Micron* **31**: 355–362.

# 8 Advanced EELS techniques in the TEM

In the previous chapters we have outlined some of the fundamental procedures for the analysis of signals in the EEL spectrum and how these may be also used for the purposes of imaging. In this last chapter we review some of the more specialized techniques that can be performed in the environment of the TEM which are of particular relevance for materials science and solid state physics.

## 8.1 Orientation dependency in EELS

### 8.1.1 *Orientation-dependent ELNES measurements of anisotropic materials*

Orientation dependency in EELS, and particularly ELNES and EXELFS, arises from any anisotropy in the atomic structure. This anisotropy leads to a directional dependence in the density of unoccupied electronic states and therefore the transitions to these final states. In the case of an isotropic material, such as an amorphous solid or a cubic crystalline material, there is no directional dependence; however, for the case of an anisotropic crystalline material, the exact form of the EEL spectrum will be influenced by the direction of the momentum transfer, $q$, relative to the unit cell axes of the material; these effects may be used to study the presence of any directionality in bonding character within the structure.

In Section 2.4 and *Figure 2.2*, we noted that it is possible to decompose the momentum transfer, $\mathbf{q}$, suffered by the incident electron during inelastic scattering into a component $q_{||}$ (parallel to $\mathbf{k_0}$ – the incident beam direction) and $q_\perp$ (perpendicular to $\mathbf{k_0}$) where $q^2 = q_{||}^2 + q_\perp^2$. For a small scattering angle, $\theta$, these components are given by $q_{||} = k_0\theta_E$ and $q_\perp \approx k_0\theta$ where $\theta_E$ is the characteristic scattering angle for the particular inelastic event. By suitable choice of the range of scattering angles contributing to the spectrum, as defined by the size of collection angle $\beta$, as well as the orientation of a single crystal specimen region, it is possible to vary the orientation of $\mathbf{q}$ relative to the crystalline coordinate system (the unit cell axes), and hence the relative weighting of these two parallel and perpendicular components of the momentum transfer. It is clear from Equations (6.3) and (6.4) that this will allow us to investigate electron transitions along certain specific directions within the crystalline unit cell. This may then be used to probe directional bond character and bond lengths by orientation-dependent ELNES and EXELFS measurements.

Experimentally, there are a number of techniques for the extraction of orientation-dependent ELNES (Botton and Boothroyd, 1995). In the CTEM, the simplest method is to use near-parallel illumination and suitably orient the sample with respect to the electron beam; use of a small collection angle (typically $< \theta_E$) then isolates principally the parallel component of the momentum transfer relative to the incident beam direction. Other methods involve ELNES measurements at two different collection angles followed by subsequent retrieval of the two individual parallel and perpendicular components. Finally, if the sample is oriented with, say, the c-axis at 45° to the electron beam, then in the STEM it is possible to measure the two separate components by displacing a small collector aperture.

*Figure 8.1* provides an example of orientation-dependent measurements of the B K-ELNES in $TiB_2$, a hexagonal material with trigonal sheets of boron and titanium arranged in alternating layers stacked perpendicular to the unit cell c-axis. The B K-ELNES measured with $q$ parallel and perpendicular to the $c$ axis exhibit significant differences,

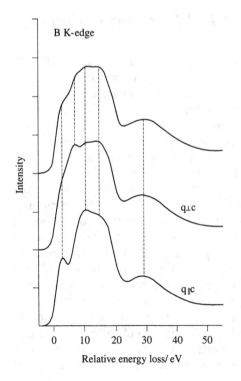

**Figure 8.1.** A comparison of the B K-ELNES of hexagonal $TiB_2$ measured with the incident beam along three different directions and a small collection angle so as to extract the parallel component of the momentum transfer: $q$ parallel to the c-axis of the crystal; $q$ perpendicular to $c$; and a general direction $q$ parallel to [2423]. The change in ELNES is due to the anisotropic nature of the titanium diboride crystal structure.

which are related to the presence of metallic bonding within the boron sheets and the existence of covalent B–Ti bonding between the layers. Modelling of these spectra with both band structure and multiple scattering calculations have also revealed the differing effects of the core hole parallel and perpendicular to the c-axis. This effect is related to the degree of metallic character and hence screening of the core hole as well as the different nature of the final states predominant in each of the two directions (Lie *et al.*, 2000).

An additional consequence of the orientation of the momentum transfer relative to the unit cell axes of an anisotropic material is that at a certain value of the collection angle (the 'magic angle', $\beta_{magic}$), the weighting of the parallel and perpendicular components will be equal and any dependence of the ELNES on sample orientation will vanish. EELS data acquired under these conditions potentially allow a much more quantitative analysis of the ELNES and, in recent years, a number of values for the magic angle have been proposed (Daniels *et al.*, 2001).

## 8.1.2 Dispersion measurements

Inspection of *Figure 2.3*, which even for ionization of a simple hydrogenic model of the atom, reveals the complex dependency between the energy loss and the momentum transfer for an inelastic scattering process, known as the *dispersion*. Use of a two-dimensional detector and/or the displacement of a small collection aperture away from the optic axis allows the investigation of the form of the dispersion for a given energy loss feature. A full dataset known as a $\omega$–$q$ plot can be recorded in one acquisition on a two-dimensional CCD array. This can be of use in determining the character of certain energy loss processes.

A volume plasmon often shows a characteristic parabolic free-electron shape for the dispersion ($E$ vs $q$) curve; however when the momentum transfer exceeds a certain value this collective oscillation cannot exist and decays into single electron excitations. Inner-shell ionization, meanwhile, is forward-peaked for energies just above the inner-shell binding energy, but exhibits a Bethe ridge at high energy loss (*Figure 2.3*) which corresponds to hard collisions with a 'free' electron. Section 8.3 discusses how measurement of the Bethe ridge can give information on the momentum densities of electrons in the solid in the particular crystal orientation under study.

## 8.1.3 Electron channelling

A further orientation dependence in EELS arises from *channelling effects*. From Section 1.2.1, solution of the Schrödinger Equation for an electron in a periodic lattice potential results in electron wavefunctions, known as *Bloch waves*, whose amplitudes are modulated by this potential. For a simple case, at an exact Bragg orientation corresponding to diffraction from a particular set atomic planes, two standing Bloch waves are set up – one with maximum amplitude peaked directly on the atomic planes and another with the maximum amplitude peaked mid-way between the atomic planes. As the specimen is tilted through the Bragg condition, the relative intensities of these Bloch waves will vary. This variation will affect the probability of inner-shell ionization (and hence the relative intensity of an ionization edge in an EELS spectrum) depending on the location within the unit cell of the atoms being ionized, relative to those atoms which compose the Bragg diffracting atomic planes. As X-ray emission follows inner-shell ionization, these channelling effects will also influence the intensities in the X-ray emission (EDX) spectrum. This technique is known as *atom location by channelling-enhanced micro-analysis* (ALCHEMI) and can be employed to investigate site occupancies and site symmetries within a particular crystalline unit cell (Egerton, 1996). Although ALCHEMI experiments are reasonably common for EDX, some EELS and ELNES studies have been demonstrated for large Bragg scattering angles (Tafto and Krivanek, 1983). For the purposes of general microanalysis, such channelling effects are often unwanted and can give rise to spurious edge intensities, which is one reason to avoid major zone axis orientations of the sample during general quantitative EELS or ELNES/EXELFS measurements.

## 8.2 Spatially resolved measurements

Spatially resolved EELS measurements are commonly obtained in dedicated STEMs, making full benefit of the small probes with high current density obtainable with a cold field emission source. However, the coming years should see these techniques being more routinely applied in the new generation of conventional TEM/STEMS fitted with thermally assisted field emission sources.

In *Figure 1.7* and Section 5.7, we noted that the use of a small, well-defined collection aperture for EELS can mediate against any loss in spatial resolution due to the broadening of the incident beam during transmission through a thin specimen. Thus, under suitable experimental conditions, the ultimate spatial resolution of an EELS measurement will depend critically on both the probe size and probe current, as well as the radiation sensitivity of the material under investigation. Clearly, the smaller the probe size, the higher the spatial resolution, although generally this will be at the expense of the experimental SNR and will also lead to higher electron doses at the specimen. An additional limiting factor for the ultimate analytical resolution, most significant at lower energy losses, will be the inelastic delocalization limit, $R$, given by Equation (7.5).

### 8.2.1 Atomic column EELS

Recent developments involving high-angle (incoherent) annular dark field (so called HAADF or 'Z contrast') imaging in the STEM has opened up the exciting possibility of atomic column by atomic column microanalysis. With the incident beam oriented along a major zone axis, columnar channelling effects, similar to those described in Section 8.1.3, limit the broadening of a small STEM probe within the specimen. For a probe of Ångström dimensions, this channelling allows the formation of an atomic resolution Z-contrast image of the specimen and, if the probe is accurately placed over selected atomic columns in the image, simultaneous EELS microanalysis from single atomic columns becomes possible (Browning *et al.*, 1993). This is of particular use for the investigation of isolated non-periodic features in a specimen such as an interface between two microstructural regions (e.g. a grain or phase boundary) or a defect (e.g. a dislocation core; Batson, 1995). Although probably the ultimate in terms of spatially resolved EELS, the general applicability of such atomic column EELS will undoubtedly be limited by the sample itself, both in terms of its sensitivity to electron-beam induced damage, the specimen drift during experimental measurement and the requirement to align the beam along a major zone axis. Nevertheless the combination of improvements in objective lens aberration correction in the STEM to produce sub-Ångström probes and advances in the read-out speed of EELS parallel detection systems promise much for atomic resolution spectrum imaging methods in the future.

## 8.2.2 Spatial difference techniques

An alternative approach to performing high spatial resolution EELS measurements with larger probe sizes and correspondingly lower electron doses is to employ the so-called *spatial difference* technique. Essentially this is a difference spectrum formed by monitoring changes in EELS and, in particular, ELNES intensity with respect to changes in beam position or spatial coordinate. Generally probe sizes in CTEM, or scanned probe areas in STEM, are of the order of a few square nanometres. The spatial difference spectrum is generated by numerically determining the difference between two (or more) spectra recorded at different locations in the sample and can reveal changes in composition and bonding (Mullejans and Bruley, 1994). A schematic diagram showing the procedure for extracting the ELNES signal at a metal–ceramic interface is shown in *Figure 8.2* (the technique is discussed in more detail in the section on Spatially resolved bonding via ELNES spatial difference measurements, p. 121). Such an approach has revealed low levels of segregant at grain boundaries and defects and has been extended to monitor changes in bonding across a boundary at nanometre resolution with better than monolayer sensitivity (Brydson, 1995; Disko *et al.*, 2001). Because the 'difference' approach highlights small changes in ELNES, care has to be taken to exclude possible artefacts such as damage and instabilities in the energy axis and beam position. Such changes appear as derivative components. Most artefacts can be identified by repeating the 'on/off' measurements in a different sequence, as well as acquiring

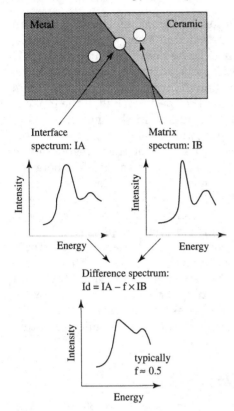

Interface
spectrum: IA

Matrix
spectrum: IB

Difference spectrum:
Id = IA − f × IB

typically
f ≈ 0.5

**Figure 8.2.** Schematic diagram of the spatial difference method for the determination of the ELNES at a metal–ceramic interface. With a beam located at points A and B, the measured intensities are IA and IB, respectively. Scaling of IB by a factor, *f*, allows modelling of the background of IA to produce the spatial difference Id.

spectra as a series of time, to ensure systematic instabilities are not a cause for concern.

***Quantification of segregation levels.*** The quantification of segregation levels at microstructural features such as interfaces and defects is becoming an increasingly important topic in materials characterization owing to the fact that trace amounts of impurities or dopants can, in many cases, modify microstructural development and affect the final mechanical, electrical or thermal properties of the macroscopic material. The accurate determination of concentrations of elemental species obtained using spatially resolved measurements in the TEM allows direct comparison of the results obtained from EELS with those derived using TEM/EDX and other techniques such as surface sensitive XPS or Auger measurements on fracture surfaces. The main benefit of EELS over EDX in the study of elemental segregation lies in higher ultimate spatial resolution as discussed in Section 5.7.

Generally the procedure for quantifying segregation levels lies in the determination of the excess concentration (in atoms per $nm^2$) of a particular element at a selected microstructural feature relative to that in the surrounding matrix; this may then be converted to a monolayer coverage. Such information may be most accurately determined in STEM mode using either EELS linescan techniques or a spatial difference method, where a scanned area of known dimensions is placed on and off the feature of interest (Brydson *et al.*, 1998). For instance, the excess concentration of an element X at a grain boundary, $\Gamma_X^{gb}$, in a matrix composed primarily of Y atoms is given by

$$\Gamma_X^{gb} = \{[n_X^{gb}/n_Y^{gb}] - [n_X^{matrix}/n_Y^{matrix}]\}.N_Y^{matrix}.w \qquad (8.1)$$

where $[n_X^{gb}/n_Y^{gb}]$ is the relative concentration ratio of elements X and Y at the grain boundary and $[n_X^{matrix}/n_Y^{matrix}]$ is the corresponding relative concentration ratio for the matrix; both these quantities can be determined by quantifying the EEL spectra using the procedures outlined in Chapter 5. $N_Y^{matrix}$ is the number of Y atoms per unit volume in the matrix, which is known from the structural unit cell data, and $w$ is the size of the irradiated area *perpendicular* to the grain boundary (taking into account any effects of beam broadening). If $N_Y$ is expressed in atoms $nm^{-3}$ and $w$ is in nm, this excess concentration is expressed in units of atoms $nm^{-2}$ and can then be converted to a monolayer coverage of element X in the Y matrix via division by a factor $(N_Y^{matrix})^{2/3}$.

The sensitivity of the spatial difference approach has been shown to be of the order of 0.1 monolayers for the determination of nitrogen segregation at platelet defects in diamond (Bruley, 1992), but this will depend critically on the probe size and current density used for analysis as well as accurate alignment of the interface or defect parallel to the electron beam.

An alternative procedure is to take an EELS linescan, that is, a set of spectra obtained by progressively moving the irradiated area across the feature of interest. This dataset can be analysed using a multivariate statistical package as outlined in Section 8.2.3, or more simply by quantifying the individual spectra and plotting the relative concentrations as a function of probe position. Fitting a simple one-dimensional Gaussian to such linescan data, and measuring the area under the curve, allows the results to be again expressed as an excess concentration or equivalently monolayer coverage (Brydson *et al.*, 1998). As discussed in Chapter 7, quantitative mapping using EFTEM is also an important technique that can additionally provide an indication of the uniformity of segregation along, for example, a particular interface. Generally for segregation studies, current results have demonstrated a resolution in EFTEM maps of around 1 nm with a detection sensitivity of about one monolayer.

*Spatially resolved bonding via ELNES spatial difference measurements.* A rapidly expanding area in EELS which utilizes the high spatial resolution attainable in TEM and STEM is the investigation of bonding at non-periodic features such as interfaces and defects (Brydson, 1995). The importance of these studies in modernday materials analysis cannot be over-emphasized as such features can often be property-determining on the macroscopic scale, whether it be in terms of overall mechanical, thermal or electronic properties. Clearly the elucidation of the bonding and localized electronic structure at such features is of great interest (Muller, 1998) and spatially resolved ELNES measurements can complement and often inform high-resolution structural imaging data, obtained using either phase-contrast imaging in HRTEM or HAADF imaging in STEM, for the determination of the atomistic arrangement and bonding arrangements at such non-periodic features unamenable to conventional diffraction techniques.

As before (*Figure 8.2*), an ELNES spectrum is taken on and off the feature of interest, for example, an homophase interface aligned edge-on to the incident beam direction. Because of the finite analysis volume, the 'interface' spectrum will actually consist of independent localized signals from both interfacial atomic species as well as those in the bulk matrix. To isolate the contribution from the interface, $I_{\text{interface}}$, the matrix spectrum, $I_{\text{matrix}}$, is normalized and subtracted from the 'interface' spectrum, $I_{\text{interface+matrix}}$, to give the spatial difference spectrum:

$$I_{\text{interface}} = I_{\text{interface+matrix}} - f \cdot I_{\text{matrix}}. \tag{8.2}$$

Procedures for determining the normalization factor, $f$, based on the analysis volume are given in Mullejans and Bruley (1994). For a heterophase interface, it may be necessary to subtract two different matrix contributions depending on the energy loss region of interest. The spatial difference spectrum represents the electronic environment of atoms at or close to the interface and may be then modelled using the techniques discussed in Section 6.2.1.

An illustrative example of the ELNES spatial difference method is provided by studies of model Cu/$\alpha$-alumina heterophase interfaces grown by MBE (Scheu *et al.*, 1998). HRTEM imaging had determined the interfacial orientation; however, image simulations were not able to distinguish unambiguously between the different possible chemistries at the interface: essentially whether there was Cu–O bonding or Cu–Al bonding between the basal plane of alumina and the (111) plane of Cu. EELS spatial difference measurements in a dedicated STEM were made using a reduced area scan, typically $3 \times 4$ nm$^2$, positioned both on the interface and in the neighbouring matrix (typically 10 nm away in a region of comparable thickness); use of a scanned area reduces the possibility of radiation damage and specimen drift affecting the measurement. The Al $L_{2,3}$, Cu $L_{2,3}$ and O $K$-edge spatial difference spectra from this interface are shown in *Figure 8.3*. All ELNES spatial difference spectra represent the contribution from interfacial atomic environments, which are substantially different from those in the neighbouring bulk matrix material (e.g. in this case, either face-centred cubic Cu or $\alpha$-alumina). This manifests itself as differences in local coordination or valency of atoms at or near to the interface. *Figure 8.3* reveals a zero Al $L_{2,3}$ spatial difference spectrum, implying that the local coordination of Al at or near the interface remains predominantly

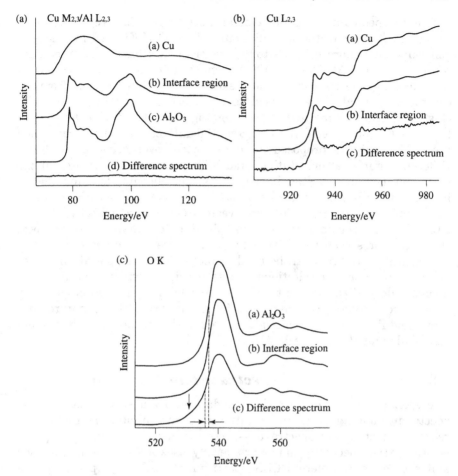

**Figure 8.3.** A set of spatial difference spectra taken from a cross sectional sample of an MBE-grown thin copper film on the basal plane of an alumina single crystal. Shown are energy regions containing: (a) the Cu $M_{2,3}$- and Al $L_{2,3}$-edges; (b) the Cu $L_{2,3}$-edge and (c) the O $K$-edge. In each case the difference spectrum is obtained from the interface spectrum by removing the contribution(s) from the neighbouring bulk material(s). See text for further discussion.

octahedral as in bulk alumina, that is, the octahedral Al $L_{2,3}$ coordination fingerprint remains predominantly unchanged. The Cu $L_{2,3}$-ELNES reveal an $L_3$ white-line feature, absent in the metal, at an energy corresponding to that observed in $Cu_2O$ – implying the existence of $Cu^+$, i.e. oxidized copper, at the interface. These results both imply the existence of Cu–O bonding at this interface, indicating, in this case, that the alumina basal plane is terminated by a layer of oxygen anions.

This chemical information provided by EELS was then used as input to HRTEM image simulations, which could now be used to adequately refine an atomistic model for the interfacial atomic arrangement. Using this model, the different interfacial environments of oxygen atoms (fifteen in total!) were calculated and used to simulate the O $K$-ELNES spatial difference spectrum using MS cluster calculations. As previously mentioned in Chapter 6, the overall shape of the O $K$-ELNES in oxides is generally much more sensitive to the longer range atomic environment, in contrast to the Al $L_{2,3}$-ELNES which exhibits a local coordination fingerprint, hence up to eight shells extending to around 0.5 nm from the central oxygen atom were used to construct the separate interfacial clusters for the MS method. Satisfactory agreement between the simulations and the measured O $K$-ELNES spatial difference spectrum confirmed the validity of the proposed atomistic model over other possibilities. Different interfacial orientations and different processing routes in the same metal–ceramic system have recently been shown to lead to different interfacial bonding mechanisms which are also detectable by EELS. This work demonstrates the synergy between HRTEM and spatially resolved ELNES measurements for the combined determination of the local chemistry and structure associated with interfaces and defects.

Identical procedures can be used to record low loss spatial difference spectra; however, two additional points require consideration: first, the inelastic delocalization limit is larger and often of the order of a few nanometres. Second, the interpretation of low loss spectra may require some modification to the classical dielectric theory described in Section 4.2 (Walls and Howie, 1989).

### 8.2.3 *Full analysis of spectrum imaging data*

The spatial difference method is fundamentally a simple form of the spectrum imaging procedures introduced in Chapter 7. Instead of recording just a few spectra or spectral regions as a function probe position, true spectrum imaging generally uses a small focussed probe and records a complete EELS spectrum at each $(x,y)$ point on the specimen and the dataset can be displayed as either as one-dimensional line profiles or two-dimensional maps.

The large datasets produced by spectrum imaging often require detailed processing to extract thickness independent spectra capable of providing information on the spatial dependence of, say, elemental compositions and/or the intensity of low loss or ELNES features. Since, in particular, the inner-shell ionization process is highly localized in a solid, each ionization event occurs independently. Thus the spectrum imaging dataset is a series of linearly superimposed signals representing the different atomic compositions and/or bonding environments within the analysis volume, and the application of statistical tools such as multivariate statistical analysis (MSA) or neural networks are ideally suited to the extraction of these individual ELNES signals (Titchmarsh, 1999). Many routines for processing spectrum imaging datasets are now becoming widely available as freeware (LISPIX).

MSA methods represent a dataset of $n$ spectra as an algebraic matrix, which can be transformed into a set of $n$ eigenvectors, each with a unique eigenvalue. The set of $n$ eigenvectors can then be separated into those $m$

components which are statistically significant and those $n-m$ which reflect noise in the data. Each spectrum in the original data set can then be decomposed into the $m$ principal components, each of which simply represent variations in ELNES or overall ionization edge intensity relative to a *'mean'* spectrum. These principle components can then be transformed into a physically interpretable data set or spectrum. The spatial difference technique outlined in the previous section can essentially be regarded as reduced dataset suitable for MSA analysis, consisting of two or possibly three spectra.

## 8.3 Electron Compton scattering

So far we have confined ourselves to the measurement of EELS data recorded at small scattering angles, typically 20 mrad or 1°. However, if electrons that have been scattered through an angle of about 5° are collected by displacing either the collection aperture or the transmitted beam from the optical axis, a new feature appears at high energy loss, well above the threshold of the highest energy ionization edge. This broad peak, shown schematically in *Figure 2.1*, is known as the electron Compton profile – corresponding to the *electron Compton scattering of solids* (ECOSS; Egerton, 1996). The origin of this feature may be seen in *Figure 2.3*, where the hydrogenic GOS is concentrated in a feature known as the Bethe ridge at high energy loss and large momentum transfer, $q$ (i.e. large scattering angle). In this regime the binding energies of the various inner-shell electrons in the solid are negligible compared to the energy loss and we have what can be approximated as a 'hard sphere' collision between two free electrons.

The energy position of the Compton peak depends on the square of the scattering angle and would be a delta function (i.e. a sharp peak) if the scattering were from stationary free electrons. However the atomic electrons in a solid are not stationary and possess a range of momenta closely related to the various electron wavefunctions and hence the bonding in the solid. This leads to a Doppler broadening of the Compton peak. Inner-shell electrons contribute mainly to the tails of the profile, whilst the central region corresponds to scattering from the valence (bonding) electrons.

Quantitative analysis of Compton data is somewhat similar to Fourier method of EXELFS analysis. After background subtraction under the Compton peak, the data is converted to a momentum scale and Fourier transformed to give the *reciprocal form factor*, which corresponds to the overlap of the wavefunction with itself (known as the auto-correlation function) in the direction of scattering (the orientation of the scattering vector). The sensitivity of the Compton profile to changes in bonding is demonstrated in *Figure 8.4* where the ECOSS reciprocal form factors for amorphous carbon, diamond and graphite are shown. Comparison of the

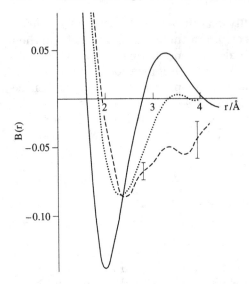

**Figure 8.4.** Fourier transformed Compton profiles for diamond (solid line), graphite (dotted line) and amorphous carbon (chained line) showing that the bonding in amorphous carbon is predominantly graphitic.

curves reveals that the bonding in amorphous carbon is predominantly graphitic (Williams *et al.*, 1984).

Since the scattering angles are large, early ECOSS studies suffered from extremely low count rates as well as problems due to multiple elastic–inelastic scattering making the technique somewhat marginal. However, improvements in detection systems and specimen preparation techniques may make this technique increase in importance as a complement to X-ray and γ-ray Compton measurements which lack the spatial resolution associated with electron spectroscopy in the electron microscope.

## 8.4 Reflection mode and surface measurements

Orienting a specimen surface near parallel to the incident electron beam allows collection of the EELS spectrum at grazing incidence measured in reflection mode geometry. Such spectra will have enhanced surface plasmon features (Section 4.2), particularly for lower incident beam energies, and may exhibit different edge fine structures. Elemental quantification procedures are similar to those presented in Section 5.4 with a modification to account for the mean distance of electrons travelling along the surface (Wang and Egerton, 1988). Since the reflection mode signal is so dependent on surface chemistry, experiments are best undertaken under ultra high vacuum (UHV) conditions or with *in situ* cleaning facilities such as specimen heating or ion beam milling.

A further related technique is measurement of EELS spectra using an 'aloof beam'; here a small probe is focused and positioned parallel to the surface of a crystal, electrons then pass close to the surface but remain in the vacuum external to the solid. The inelastic delocalization limit means that even in this configuration, it is still possible to excite electronic transitions in the solid; generally bulk volume plasmons are not excited and surface excitations alone can be studied (Howie, 1983).

# References and further reading

**Batson, P.E.** (1995) Near atomic resolution EELS in silicon-germanium alloys. *J. Microsc.* **180:** 204–210.

**Botton, G.A. and Boothroyd, C.B.** (1995) Momentum dependent energy loss near edge structures using a CTEM: the reliability of the methods available. *Ultramicroscopy* **59:** 93–107.

**Browning, N.D., Chisholm, M.F. and Pennycook, S.J.** (1993) Atomic resolution chemical analysis using a STEM. *Nature* **366:** 143–146.

**Bruley, J.** (1992) Detection of nitrogen at {100} platelets in a type IaA/B diamond. *Phil. Mag. Lett.* **66:** 47–56.

**Brydson, R.** (1995) Probing the local structure and bonding at interfaces and defects using EELS in the TEM. *J. Microsc.* **180:** 238–249.

**Brydson, R., Chen, S., Pan, X., Riley, F.L., Milne, S.J. and Ruhle, M.** (1998) Microstructure and chemistry of intergranular glassy films in liquid phase sintered alumina. *J. Am. Ceram. Soc.* **81:** 396–379.

**Daniels, H., Brown, A., Scott, A., Nichells, A., Rand, B. and Brydson, R.** (2001) Experimental and theoretical evidence for the magic angle in transmission electron energy loss spectroscopy. *Ultramicroscopy,* submitted. Proceedings EMAG 2001, Institute of Physics Conference Senes (2001). IOP Publishing, Bristol.

**Disko, M.M., Ahn, C.C. and Fultz, B.** (2001) *Transmission Electron Energy Loss Spectrometry in Materials Science,* 2nd Edn. TMS, Warrendale, Pennsylvania.

**Egerton, R.F.** (1996) *Electron Energy Loss Spectroscopy in the Electron Microscope.* Plenum Press, New York.

**Howie, A.** (1983) Surface reactions and excitations. *Ultramicroscopy* **11:** 141–148.

**Lie, K., Hoier, R. and Brydson, R.** (2000) Theoretical site- and symmetry resolved DOS and experimental EELS near edge spectra (B $K$-ELNES) of $AlB_2$ and $TiB_2$. *Phys. Rev. B* **61:** 1786–1794.

**LISPIX** – a public domain scientific image analysis program for the PC and Macintosh. *http://www.nist.gov.lispix/*

**Mullejans, H. and Bruley, J.** (1994) Improvements in detection sensitivity by spatial difference electron energy loss spectroscopy at interfaces in ceramics. *Ultramicroscopy* **53:** 351–360.

**Muller, D.A.** (1998) A simple model for relating EELS and XAS spectra of metals to changes in cohesive energy. *Phys. Rev. B* **58:** 5989–5995.

**Scheu, C., Dehm, G., Mobus, G., Ruhle, M. and Brydson, R.** (1998) Electron energy loss spectroscopic studies of copper/alumina interfaces grown by MBE. *Phil. Mag. A* **78:** 439–465.

**Tafto, J. and Krivanek, O.L.** (1983) Site specific valence determination by EELS. *Phys. Rev. Lett.* **48:** 560.

**Titchmarsh, J.M.** (1999) Detection of electron energy loss edge shifts and fine structure variations at grain boundaries and interfaces. *Ultramicroscopy* **78:** 241–250.

**Walls, M.G. and Howie, A.** (1989) Dielectric theory of localized valence energy loss spectroscopy. *Ultramicroscopy* **28:** 40.

**Wang, Z.L. and Egerton, R.F.** (1988) Absolute determination of surface atomic concentration by reflection electron energy loss spectroscopy. *Surf. Sci.* **205:** 25–37.

**Williams, B.G., Sparrow, T.G. and Egerton, R.F.** (1984) Electron Compton scattering from solids. *Proc R. Soc. Lond.* **A393:** 409–422.

# 9 Conclusions

In this short review I have attempted to indicate the range of both qualitative and quantitative analytical techniques associated with transmission EELS. These include methods for determining a large array of solid-state properties, including the distribution of light elements, the sample thickness, the crystallographic structure, the bonding characteristics and the electronic structure.

The use of EELS in the environment of the TEM allows these techniques to be conducted at high spatial resolution and provide some measure of the physical, chemical and microstructural properties of the material under investigation. This field of research is rapidly expanding, as information on nanostructure–property–processing relationships becomes increasing valuable in the modelling, development and tailoring of advanced materials for specific applications in nanoscale science and technology.

Although the electron spectroscopic techniques have been described principally from the standpoint of the physical sciences, it is clear that the techniques have considerable potential in both the geological sciences and life sciences, particularly in the interface of the latter with the physical sciences. Notable techniques here include the high spatial resolution mapping of light elements and the determination of valence states in both inorganic and organic structures.

Below we list a selection of useful resource material, which will provide a basis for further research in the subject.

## 9.1 Further reading

### 9.1.1 Monographs

**Ahn, C.C. and Krivanek, O.L.** (1983) *EELS Atlas*. ASU Centre for Solid State Science, Tempe, Arizona and Gatan Inc., Warrendale, Pennsylvania.

**Disko, M.M., Ahn, C.C. and Fultz, B.** (2001) *Transmission Electron Energy Loss Spectrometry in Materials Science*, 2nd Edn. TMS, Warrendale, Pennsylvania.

**Egerton, R.F.** (1996) *Electron Energy Loss Spectroscopy in the Electron Microscope*. Plenum Press, New York.

**Fuggle, J.C. and Inglesfield, J.E. (eds)** (1992) *Unoccupied Electronic States. Fundamentals for XANES, EELS, IPS and BIS*. Springer, Berlin.

**Keast, V.J., Brydson, R., Williams, D.B. and Bruley, J.** (2001) Electron energy loss near-edge structure – a tool for the investigation of electronic structure at the nanometre scale. *J. Microsc.* (in press).

**Raether, H.** (1980) Excitations of plasmons and interband transitions by electrons. *Springer Tracts in Modern Physics*, Volume 88. Springer, New York.

**Reimer, L. (ed.)** (1995) *Energy Filtering Transmission Electron Microscopy*. Springer, Heidelberg.

**Stohr, J.** (1992) *NEXAFS Spectroscopy*. Springer, Berlin.

**Williams, D.B. and Carter, C.B.** (1997) *Transmission Electron Microscopy*. Plenum Press, New York.

## 9.1.2 Conference proceedings

### Proceedings of dedicated EELS workshops

Lake Tahoe, August 1990 – *Microsc. Microanal. Microstruct.* **2:** 143–412 (1991).

Leukerbad, July 1994 – *Microsc. Microanal. Microstruct.* **6:** 1–158 (1995) and *Ultramicroscopy* **59:** 1–292 (1995).

Port Ludlow, 1998 – *Micron* **30:** 101–194 (1999) and *Ultramicroscopy* **78:** 1–274 (1999).

### EELS sessions in Microscopy Conferences

EMAG–Institute of Physics UK: *Institute of Physics Conference Series*, Nos. 161 (1999); 157 (1997); 147 (1995); 138 (1993); 119 (1991). IOP Publishing, Bristol.

Microscopy Society of America (MSA): *Microscopy and Microanalysis Series*, Vols 1–6. Springer, Berlin.

International Congress for Electron Microscopy (ICEM): *Electron Microscopy* 1998 (IOP Publishing); 1994, 1990 (San Francisco Press).

European Congress for Electron Microscopy: *Proceedings of EUREM-11* 1996 (CESM: Brussels 1998); *Electron Microscopy* 1992 (University of Granada).

---

# 9.2 Further resources

## 9.2.1 Research

EELS and X-ray Database: CEMES-LOE/CNRS Toulouse, France;
  *http://www.cemes.fr/~eelsdb/*
Microscopy and Microanalysis WWW server: Argonne National Laboratory;
  *http://www.amc.anl.gov/*
Daresbury Crystal Structures database: Daresbury Laboratory;
  *http://cds3.dl.ac.uk/cds/cds.html/*
Crystal Lattice Structures; *http://est-www.nrl.navy.mil/lattice/*
WebElements: Argonne National Laboratory; *http://www.webelements.com/*

## 9.2.2 List servers/mailing lists/news groups

Royal Microscopical Society; *http://www.rms.org.uk/*
Institute of Physics, Electron Microscopy and Analysis Group (EMAG);
  *http://www.iop.org/IOP/Groups/EM/*
LEMAS listserver; *http://www.lemas.co.uk/* and
  *http://www.jiscmail.ac.uk/lists/lemas.html/*
Electron Microscopy Yellow Pages, Centre for Electron Microscopy of the Ecole Polytechnique Federale de Lausanne; *http://cimewww.epfl.ch/emyp/*

Micro World, Internet guide to Microscopy and related sites;
  *http://www.mwrn.com/guide.htm/*
Microscopy and Imaging Resources on the WWW, Experimental Pathology Service Core, Tucson, University of Arizona;
  *http://www.pharm.arizona.edu/centers/tox_center/swehsc/exp_path/m-I_onw3.html/*
Computer Network Laboratory for Microscopy Education, Materials Science and Engineering Department, Tucson, University of Arizona;
  *http://aluminium.sem.arizona.edu:8001/*
National Center for Electron Microscopy, Laurence Berkeley Laboratories, University Of California; *http://ncem.lbl.gov/ncem.html/*
Microscopy Online, Microscopy related Newsgroups and Listservers;
  *http://www.Microscopy-Online.com*

## 9.2.3 Training and education

The Electron Microscopy Outreach Program, San Diego Supercomputer Centre;
  *http://em-outreach.sdsc.edu/EM_Outreach.html/*
Materials Science on CD-ROM, Chapman and Hall, London (1998);
  *http://www.matter.org.uk/*

## 9.2.4 Software tools

### Modelling and simulation
Crystal Maker & Crystal Diffract for Macintosh; *http://www.crystalmaker.co.uk/*
Crystal Kit and Mac Tempas (Imaging/Diffraction calculations); *http://www.totalresolution.com/*
Cerius$^2$ – various modules covering a broad range of simulation and modelling;
  *http://www.msi.com/*
FEFF8 – multiple scattering modelling of ELNES;
  *http://leonardo.phys.washington.edu/feff/*
WIEN97 – band structure (FLAPW) modelling;
  *http://www.tuwien.ac.at/theochem/wien97/*
Desktop Microscopist; *http://www.easystreet.com/ ~lacuna/*

### Image acquisition and processing
EELS spectrum processing (ELP) software; *http://www.gatan.com/software/*
MSA public domain software library; *http://www.amc.anl.gov/*
NIH Image – Public Domain Software; *http://rsb.info.nih.gov/nih-image/*
Image Processing Tool Kit; *http://members.aol.com/ImagProcTK/index.htm/*
Digital Micrograph; *http://www.gatan.com/software/DigitalMicrograph.html/*
Tietz Video and Image Processing Systems; *http://www.tvips.com/*
Spectrum Imaging software; *http://www.emispec.com/* and *http://www.gatan.com/*
Multivariate Statistical Analaysis software; *http://www.nist.gov.lispix/*

## 9.2.5 Product databases

Gatan; *http://www.gatan.com/*
LEO; *http://www.leo-em.co.uk/*
Electron Microscopy Software; *http://www.kaker.com/mvd/products/soft.html/*
Image Analysis and Processing; *http://www.kaker.com/mvd/products/image.html/*
Microscopy Vendors Database; *http://www.kaker.com/mvd/vendors.html/*

# Index

Printed in the United States
by Baker & Taylor Publisher Services